Mining the Sky

Mining THE Sky

UNTOLD RICHES FROM THE ASTEROIDS, COMETS, AND PLANETS

JOHN S. LEWIS

HELIX BOOKS

BASIC BOOKS

A Member of the Perseus Books Group
New York

Library of Congress Cataloging-in-Publication Data
Lewis, John S.
   Mining the sky : untold riches from the asteroids, comets, and
planets / John S. Lewis.
      p. cm.
   Includes bibliographical references and index.
   ISBN 0-201-47959-1
   ISBN 0-201-32819-4 (pbk.)
   1. Outer space—Exploration—Research—Government policy—United
States. 2. Outer space—Exploration—Research—Industrial
applications—Government policy. I. Title
QB500.266.U6L49   1996
333.9'4—dc20
                                                          96-15927
                                                          CIP

Cover design by Suzanne Heiser
Text design by Diane Levy
Set in 12-point Bodoni by Carol Woolverton Studio

3 4 5 6 7 8 9 10

Basic Books is a Member of the Perseus Books Group.

Find Basic Books on the World Wide Web at
http://www.basicbooks.com

To Van, Meg, Chris, Katie, Bonnie, and Peter
and
for a future of limitless opportunity, joy, and growth
on Earth
as it is in the Heavens

Destiny is not a matter of chance—
it is a matter of choice.
It is not a thing to be waited for—
it is a thing to be achieved.

—William Jennings Bryan

CONTENTS

PREFACE

Many scientists earnestly anticipate a vigorous resumption of pure scientific research on the solar system. As much as I would like to see the resumption of space exploration, I find it quite incredible that any nation on Earth would choose to devote substantial resources to such an endeavor for its intellectual value alone. As a rule, governments are not intellectually inclined. If we are to return, for example, to the Moon, it will be because there is some visible relationship between that endeavor and the future material well-being of our nation and planet. Basic research will be tolerated by the political machinery only if it constitutes a *balanced part of a research program that also satisfies visible economic needs.*

There are many who cast the debate in polar terms. Some argue that the government must support basic research because it is the basis of the future; that all the new applied science of a decade hence and all the new engineering developments of twenty years hence will build on today's basic science; that American basic science leads the world and must not be allowed to falter. These points are true. The problem is that this message has not been delivered with the necessary explanations and case studies to those in political power. The scientists whose ox was gored are so "pure" that they don't know—or honestly don't care—about applied science and the commercialization process.

Most politicians, on the other hand, see basic research from the perspective of American industry, which has become increasingly obsessed with the current quarter's balance sheet. Many American industries see that basic research is a major expense this quarter, and that this year's basic research will not result in a single new product this year. They therefore have chosen to "improve" their ledgers by cutting funding for basic research. That such behavior is suicidal in the long run is not at all obvious to them, nor does it address their short-term needs.

The debate between these parties, predictably, usually devolves into impassioned exchanges between basic scientists who say, in effect, "We're very important and you aren't smart enough to understand why," and politicians, who say, "We need research that results in short-term economic benefits, and you're not doing it." It is one short step from there to name-calling, with scientists lambasting "know-nothing" politicians for their "arbitrary, draconic" policies and politicians lambasting "ivory-tower" scientists and their "entitlement programs."

This is the climate within which decisions about the future of space research and resource use are being made. This book is addressed to those intelligent enough to grasp that both sides of this debate are raising fundamentally correct and important points, and that the unconditional victory of either of these camps would be bad for the nation. We henceforth assume a future policy in which a judicious balance between long-term basic research, short-term applied research, engineering development of products, and commercialization of new products has been reached. Any Japanese industrialist would approve of this scheme; indeed, decades ago Japan identified the strength of American basic research and the weakness of Japanese basic research as a problem affecting Japan's future competitiveness. Funding for basic research in Japan has grown dramatically in recent years to compensate for this imbalance, but that research is integrated with and responsive to the *long-term* needs of industry. Contemporary Japanese critics of the American system would identify as its principal weaknesses not the amount of money lavished on basic research, but the poor linkage between basic and applied research and the achingly slow commercialization process of American industry, much of which is structured to resist change, not encourage innovation. The most astute critic might see the single cause at the root of these difficulties: the ascension into top management of a generation of managers whose education trains them to count beans but leaves them perfectly ignorant of where future crops of beans come from. To a large extent, the solution lies in the same place. In a very real sense, scientifically and technically literate MBAs could save American industry.

The current budget debates in Washington take place under the assumption of limited resources. The spirit of the time seems to be that things are pretty bad and will certainly get worse. Liberals tell us we are running out of natural resources and cannot use the ones we have because energy production, mining, and industry pollute. Conservatives tell us that we are running out of money and can revive the economy only by slashing funding for research and education and lowering environmental standards. The message of this book will not sit well with either camp.

The truth is that the resources available to us are, for all practical purposes, infinite. Building on what we know of the solar system, and using presently available or readily foreseeable technologies, we can relieve Earth of its energy problem, make astronomical amounts of raw materials available, and raise the living standard of people worldwide. We only need to lift up our eyes and look at the wealth of energy and materials that surrounds us in space. That vision will inspire us to seek out ways to make economical use of them.

ACKNOWLEDGMENTS

I am grateful to my wife, Ruth A. Lewis, for her careful reading of the manuscript and for her relentlessly helpful suggestions of changes and improvements.

I am indebted to David Egge for the cover illustration, to him and William K. Hartmann for permission to reproduce several paintings, and to Steven Ostro for the use of the radar image of the asteroid Toutatis.

1

INTRODUCTION:
THE END IS NEAR?

The terrestrial archives of the late twentieth century are filled with strange incongruities. For example, in news sheets printed on vegetable pulp from their years 1970 to 1979 (in their ten-based system, roughly corresponding to the years 4F26F to 4F272 in our calendar), there are many humorously intended drawings portraying old, bearded men wearing sandals and long robes, carrying signs that say, in effect, "The end is near!" The sandals and robes in these "cartoons" were meant, in the context of that time and place, to convey the sense that such men, called prophets, were anachronistic remnants of an ancient and obsolete tradition, eccentric but harmless. The sophisticated urbanites of that time prided themselves on a "scientific, enlightened" worldview that rejected catastrophic or apocalyptic predictions, especially those with religious or philosophical overtones. Science was then clearly in public favor; several expeditions had recently been successfully dispatched to their Moon, a class B7p(1:1 locked) single satellite.

In bizarre counterpoint to their tone of gentle mockery of catastrophes, one of the most widely reported news topics of the decade was the development of a truly apocalyptic computer model of the terrestrial economy. Computers were then in a state of extremely rapid technological evolution, driven largely by international military and space competition, with annual dramatic increases in computer size and speed. Many problems which had been computationally intractable just a few years earlier, such as long-range weather prediction, nuclear reaction network analysis, and planet-wide economic modeling, were then under study for the first time. Thus, the same news sheets that mocked the sandaled prophets also reverently reported the work of economists from a prominent university, the Massachusetts Institute of Technology. These savants purported to show that the resources of their planet must inevitably be exhausted within a few tens of years! No matter what assumptions regarding standards of living, energy consumption, recycling, etc. were inserted into their model, the result was always global famine and economic chaos within a single human lifetime. These calculations, astonishingly, made no allowance for space technology and the access it provides to new sources of energy and materials. Their message was unambiguous: "The end is near!"

—From *Sociological and Economic Studies
of Terrestrial Archival Materials,
Overview Unit 2, 4F3A1*

⋰ ✳ ⋰

The age of exploration in which we live commenced (or so we are told) about the year 1419 with the first true European voyage of exploration, a Portuguese expedition to the Madeira Islands. Fostered by new navigation and shipbuilding technologies introduced by Prince Henry the Navigator of Portugal, European sailors opened first Africa, then India, and then the Americas and East Asia to European economic, religious, political, and military influence. A small percentage of the planet's population—principally the residents of England, Spain, France, and Portugal—came to exercise dominion over most of Earth. Smaller or less vigorous exploring nations, such as the Netherlands, Italy, Belgium, Germany, and Sweden, founded colonies but soon lost them or found them hemmed in by the leading European maritime powers. English became the language of the United States, Canada, New Zealand, and Australia, and the unifying tongue of India and many African nations. French became the common language of West Africa, and Spanish and Portuguese filled the Americas from the Rio Grande to the Antarctic.

This is our familiar textbook history; in many ways, the European domination of the world seems to have been inevitable. The trade goods, profits, and cultural stimulation of the age of exploration fueled the economic and social revitalization of Europe. Simultaneously, the sudden imposition of European ideas and religion caused a devastating cultural earthquake in conquered lands. Most of the world quickly became dominated by European culture and language. But it almost did not happen.

During the Ming dynasty a passion for exploration swept China. Starting in 1403, great treasure fleets were dispatched to map the coasts of East Asia and establish trade relations. The fleets headed westward, opening the coasts of India and Arabia to Chinese trade. Later Ming expeditions progressed down the east coast of Africa, rounded the Cape of Good Hope, and headed north. They sailed past the Congo to the great bulge of West Africa, reaching the very doorstep of Europe about the time that the young Henry the Navigator convened his great assembly of pilots, mapmakers, shipwrights, astronomers, and instrument makers at Sagres in 1421. But just before the fleets were due to break into European waters, establish trade with Portugal and Spain, and discover the entrance to the Mediterranean, word of the Chinese emperor's death reached the fleet. The court eunuchs, alarmed at the "waste" of resources on the voyages, recalled the fleet and ordered an end to exploration. There was dissent. The eunuchs ordered the treasure fleets beached and burned. There was still dissent. The eunuchs ordered the plans and records of the expeditions collected and burned. By 1433 China had again cut itself off from the rest of the world.

China undoubtedly could have become the dominant power throughout the world if it had only persevered in its program of exploration and trade for a few more years. The sole language of the modern world would have become Mandarin Chinese.

Similar choices face us in the closing days of the twentieth century. We have sent forth our exploratory fleets to map the far side of the Moon, never before seen by human eyes; to pierce the clouds of Venus and see its hellish surface; to land on Mars and seek signs of life on its surface; to sail by the giant planets Jupiter, Saturn, Uranus, and Neptune and their retinues of satellites; to explore the byways of the archipelagoes of asteroids. Twelve men from Earth have walked on the Moon. News of the fabulous wealth of energy and materials within our grasp has begun to trickle back to Earth—but the emperor is dead. There is no leadership; the dream must die. At the height of the Apollo program, the National Aeronautics and Space Administration (NASA) canceled production of the *Saturn 1* and *Saturn 5* boosters. With them, the human race lost its ability to send astronauts to the Moon and beyond. Then began the enormously expensive diversion of the Space Shuttle program, an incomparably less capable booster with which we cannot even dream of flights to the Moon. The last three flight-ready *Saturn 5* boosters, already built and paid for, were laid out as lawn ornaments at Cape Canaveral, Marshall Space Flight Center, and Johnson Space Center, to rust into ruin and serve as testimony to the futility of that dream. Only governments can afford to spend a quarter of a billion dollars each for the equivalent of plastic flamingos.

The tools and dies for building the Saturns were collected and sold as scrap metal for pennies a pound; a $20 billion investment in the future was melted into dross. The plans and blueprints and operating instructions for the Apollo hardware were declared surplus. One set of *Saturn 5* plans was donated to a Boy Scout paper drive. The last apparently complete set of plans was sent to the Federal Record Archives in Atlanta. My attempts to find them several years ago met with no success: the plans have evidently been "lost." The fleet has been destroyed. The plans are gone. The eunuchs have won the day.

.⁝.✳.

Despite the rarity of visionary leadership in recent years, Earth has been blessed with a host of true visionaries. Since the early 1900s, certain engineers, scientists, and writers of futuristic fiction have foreseen a human presence in space that is sustained not by political posturing but by economic and practical self-sufficiency. The idea that technology could make the heavens accessible and useful to us dates from the beginning of this century, from the pioneering and independent work of Konstantin Tsiolkovskii in Tsarist Russia, Robert Goddard in the United States, and Hermann Oberth in Germany (see chapter 2). We shall see how writers of science, engineering, and science fiction, such as Arthur C. Clarke, Olaf Stapledon, and J. D. Bernal, sketched out ways to make human explorers and colonists eventually self-sufficient in space. But these visions were not shared by political leaders, even after the space age had arrived. Indeed, even in the Apollo exploration of the Moon, the concept of using lunar materials was avoided because of engineering conservatism and short-sightedness: the lunar landings were regarded as a short-term, dead-end political goal, not a long-term commitment to move into space on a firm economic footing.

The United States and the Soviet Union undertook the exploration of space as part of the fierce rivalry of the cold war, to hone technologies of potential military value while demonstrating their competence in peaceful but technically challenging competition. It was that bitter rivalry that led us to lift our gaze to space—but what we have seen has transfigured us. The space age has opened our eyes both to the rare beauty and vulnerability of Earth and to the incomparable richness and variety of our cosmic surroundings. Seeking to score points against each other in the eyes of global opinion, the American and Soviet exploration programs enlisted thousands of scientists and engineers whose efforts have served to make space into a real place, ever more accessible to all humankind. They have found, from the study of other planets, new insights into the nature and governing laws of worlds. Many of these insights have been brought home to Earth, to inform us of the nature of our world, instruct us in its governing laws, and guide us in designing public policy so as to preserve our beautiful Earth. Venturing forth into space, we expected to find a harsh, deadly vacuum, a quarantine notice for the human race. Instead, we have found a lively, rich understanding of the unity and lawfulness of Creation, within which the diversity and complexity of local materials and events fall naturally into place. The deadly vacuum is still there, but now we know vastly more about the laws of living in space and about the resources available to sustain life off Earth. These resources in fact greatly surpass those available to us on Earth's surface.

The tired old myths of limited resources and finite living space available to humankind cannot be sustained in the face of the discoveries now pouring in daily about our solar system. Indeed, the global expansion of European technology and civilization brought about by the terrestrial age of exploration is but a pale foreshadowing of the opportunities before us as humans move out into space.

Sadly, the domination of space activities by governments has caused the price of space travel to remain very high; indeed, with the Space Shuttle, the cost of access to space actually rose farther, reaching about ten thousand dollars per kilogram of payload carried into orbit. Governments do not need to show a profit, and they have strong institutional incentives to keep costs (and therefore budgets) as high as possible. This tradition of high costs arising out of centralized governmental control has generated another "tired new myth"—that space travel *must* be very expensive. But this is most certainly not true. The cost of launching payloads from Earth into space can be brought down dramatically by using the most elementary principles of free enterprise, starting with free and open competition. Several launch vehicles now in development could promise launch costs under a thousand dollars per kilogram at the very beginning of the competitive era. The ultimate result of competition will be that the cost of transportation into space will be only a few times higher than the cost of the fuel. The actual cost of fuel to carry payload into space is, astonishingly, far lower than present launch costs—about $3.00 per kilogram of payload placed in orbit by a rocket fueled with liquid hydrogen and liquid oxygen! In fact, at the present price of electric power, the energy needed to put a kilogram of payload into orbit costs only about $0.50 per kilogram. A 75-kilogram (165-pound) adult would, at this price, pay $37.50 for the energy to launch himself into space. Allowing for lifting the vehicle as well as the passengers, paying salaries, and setting aside a decent profit margin, a ticket to orbit could go for as little as $500. Once in orbit, you are precisely halfway to having enough energy to escape from Earth. For twice as much energy, and twice the expense, you could escape from Earth entirely. But if you were going to a remote destination, such as the asteroid belt, the cost of your transportation would be dominated by the cost of launching the food, air, and propellant you would need to get you to your destination. Of course, if the fuel and life-support materials did *not* have to be lifted from Earth, then you could go just about anywhere in the solar system for a few thousand dollars!

Once in space, we shall probably use the resources of other solar system bodies simply to provide the consumable propellants and life-support materials to reduce the costs and increase the performance of space missions. The first body to be exploited will probably be the nearest body to us in space, our Moon. The Moon is unfortunately a barren body devoid of atmosphere and liquid water. Studies of the lunar surface by American and Soviet spacecraft and by Earth-based telescopes have revealed a slaglike composition for the lunar crust, rich in silicates and oxides, but very poor in both volatile materials (hydrogen, carbon, nitrogen, and so on) and free metals, as discussed in chapter 3.

Despite its unattractive aspects, the Moon presents several opportunities for resource use, at first mostly to defray the expense of expeditions to explore its surface. In chapter 4, we shall discuss how oxygen extracted from lunar rocks or lunar polar ice (if present) can be used for life support and as rocket propellant. Polar ice also would provide hydrogen for rocket fuel and for industrial processes. Lunar dirt can easily be used as radiation shielding, and metals can be made as a byproduct of oxygen production. The principal market for these products would be on the Moon itself. Export of propellants in support of large near-Earth space activities may conceivably be economical.

But the Moon is by no means the only accessible body for resource use. As discussed in chapter 5, Spacewatch and other astronomical search programs have found that Earth is embedded in an immense swarm of comets and asteroids that threaten us with collision and global mass destruction. Thousands of kilometer-sized bodies cross Earth's orbit. Many recently discovered asteroids that follow orbits similar to Earth's have very high probabilities of impact. There are at this moment at least four hundred Earth-crossing asteroids larger than a kilometer in diameter that *will certainly* collide with Earth. The only uncertainty is *when* these devastating impacts will occur. The very asteroids that present the greatest danger to us are also the most accessible bodies in the solar system. Many of the most dangerous are actually easier to reach and land on than the Moon! Protecting Earth from the catastrophic impact of such bodies can be achieved using known technology, as I discussed in my book *Rain of Iron and Ice.* *
But any space propulsion systems developed to give us access to these bodies for defensive reasons also gives us access to their resources. These nearby bodies contain a wide range of valuable materials: many

*Addison-Wesley, Reading, MA, 1996.

are rich in water (both as ice and water-bearing minerals). Natural stainless steel, containing many valuable metals such as nickel, cobalt, and the platinum metals, makes up 10 percent to 99 percent of the mass of most asteroids. In chapters 5 and 8 we explore ways in which fundamentally important industrial materials such as water and metals can be extracted automatically from near-Earth asteroids and extinct comets. These raw materials can be used to make propellants, life-support materials (air, water, nutrients), and structural metals in quantities large enough to make in-space operations profitable and self-sustaining. Among the commodities that might be returned for use on Earth are precious and strategic materials and cheap, clean, abundant energy. As an added benefit, industrial use of these bodies protects human civilization on Earth from the danger of destruction by their impact.

But the resources of the Moon and asteroids are of no interest unless we have economical access to them, as discussed in chapters 6 and 7. Access means that we have technically feasible, economically competitive transportation systems, with low propulsion energy requirements and opportunities for repeated round-trips. Taking advantage of propellants made on the Moon or of asteroids greatly extends the list of attractive and affordable missions to Mars, the asteroid belt, and elsewhere by dramatically lowering Earth launch costs. In chapter 7 we see how repeated round-trip missions to selected near-Earth asteroids can pay off by returning 100 times the mass originally launched from Earth. Logistic considerations make the Moon a very attractive base of operations if ice is abundant in the lunar polar regions. Export of metals and propellants from the Moon is of marginal attractiveness because of the Moon's substantial gravity.

One of the few commodities that we might hope to import to Earth from space profitably is energy. We discuss in chapter 8 how importation of energy from space can relieve our planet of the burdens of fossil-fuel combustion and nuclear-waste disposal. The 1977 proposal to launch solar power satellites into high orbits to transmit electrical power to Earth gains considerable benefit from the use of lunar or asteroidal material. We could build the low-tech parts of these power satellites (wires and cables, propellants, beams and fixtures, and maybe even solar cells) in space because launch costs from the Moon or nearby asteroids could be lowered to less than launch costs from the much more massive Earth. Alternatively, arrays of solar cells may be made on the Moon out of lunar materials and only the power exported. The lunar surface also contains the light isotope helium-3, which is implanted in the surface by the solar wind. Helium-3 and heavy hydrogen (deuterium) can in principle be used as the fuel of huge fusion reactors to generate cheap, clean electricity. Finally, if solar power satellites are built in high orbits about Earth, by far the most attractive source of raw materials to build them may be small nearby asteroids with negligible gravity.

The only other bodies besides the Moon and nearby asteroids and comets that are reasonably accessible from Earth are Mars and its two natural satellites, Phobos and Deimos. In the survey of Mars in chapter 9, we see that the red planet has bright, extensive polar caps of ice and solid carbon dioxide (dry ice). The surface is widely and deeply scored by the valleys of ancient rivers comparable in flow to the Amazon. Certain Martian surface features have been interpreted as evidence of ancient ocean beds and widespread glaciation. The craters made by large cometary and asteroidal impactors give strong evidence of a deep layer of permafrost and ground ice over most of the planetary surface. Water is extremely attractive as a source of both propellants and life-support materials. Polar ice, snow, and permafrost are widespread, and water-bearing minerals are probably universal on the Martian surface. All are testimony to a wetter, more benign past. Expeditions are made far easier and cheaper by use of rocket propellants made on the surface of Mars. In chapter 10 we explore how propellants can be readily made from carbon dioxide in the atmosphere, from water, and from other native materials. Even unpiloted sample-return missions would benefit greatly from the automated local production of propellants. The frozen present of Mars conceals a warmer, wetter past—and allows a lively future.

Phobos and Deimos are strange and wonderful bodies. Both of these two small, irregular moons are generally similar to asteroids, but they are gravitationally bound to Mars (chapter 11). If they are similar to the carbonaceous chondrite meteorites, they would be very attractive refueling stops in the Mars system, providing propellant for landings on Mars and for return from the Mars system to Earth.

The attractiveness of asteroids as resources is now obvious. But we have so far concentrated exclusively on the small minority of asteroids that have orbits bringing them close to Earth. What of the vast majority of asteroids, which orbit about the Sun far beyond the orbit of Mars, in the asteroid belt? As discussed in chapter 12, many of the near-Earth asteroids that are more accessible than the Moon pursue eccentric orbits that approach the Sun close enough to graze or cross Earth's orbit, and then they retreat as far out as the heart of the asteroid belt. Any equipment installed on these bodies will get a free trip to the belt every few years as the asteroid pursues its orbit about the Sun. It is relatively easy to ride a near-Earth asteroid out from Earth to the belt, and there to transfer to any of tens of thousands of large belt asteroids. The wealth of materials available in these asteroids, discovered in the study of meteorites, staggers the imagination. There is enough raw material in the asteroids to support a population much larger than that of Earth.

We have so far concentrated principally on the use of space resources to support future activities in space. In so doing, we have found that the energy resources of the inner solar system, including both solar power and fusion of lunar helium-3 with terrestrial deuterium, are sufficient to meet the foreseeable energy needs of Earth for the indefinite future. But power for activities in the belt and for propulsion of spacecraft is harder to find. The intensity of sunlight is about ten times lower in the belt than in the inner solar system, making all solar-power schemes less attractive. Extraction of helium-3 from the surfaces of asteroids is not likely to compare with that from the Moon. But the atmospheres of the Jovian planets contain astronomical amounts, and high concentrations, of both helium-3 and deuterium. The massive gravity wells of Jupiter and Saturn render them very unattractive sources of materials, but Uranus (and Neptune) have low enough escape velocities to be amenable to export of helium-3. The availability of this energy source allows an infinite future, which we explore in chapter 13.

The implications of abundant, cheap, clean energy for Earth are profound. We are all aware that Earth in the coming third millennium faces a variety of urgent environmental problems, and in chapter 14 we consider new solutions. We must keep in mind planet Earth's ebbing supplies of fossil fuels, emerging environmental hazards (such as global warming, acid rain, ozone depletion) caused by combustion of hydrocarbons and coal, and the uncertain prospects for safe disposal of radioactive waste from nuclear power plants. These discouraging factors combine to compel a search for cheaper and environmentally safer energy sources. Solar power and fusion power are the two leading contenders. The exploitation of space technology offers the most attractive solution to the energy problem now visible. We also should anticipate having powerful new technologies at our disposal for our use in space, drawn from fields as diverse as plasma physics, computer science, and genetic engineering. New tools from these fields will greatly diminish the cost of doing the work that needs to be done.

The expansion of human civilization into space is feasible because of the availability of vast asteroidal and planetary resources. But these resources will be ours only if we choose to pursue them. Chapter 15 lays out the choice before us: we may choose either to retreat into a stagnant, shrinking future and play a zero-sum game on an exhausted planet or to expand into a vast new arena of activity, rich in material wealth and vibrant with energy, which will allow us both the freedom to escape the cradle and the resources to keep our home planet alive and well.

2

HUMANITY
AWAKENING TO SPACE

When a man is called, he is often called young. I was nine years old when my future whispered in my ear and changed my life. Looking back from the vantage point of ninety-four years of active life, I have no doubts about when and where it happened. On January 3, 1950, my parents took me into New York to visit the Hayden Planetarium for a birthday surprise. I was interested in astronomy (and a dozen other things as well); I had read several short "guides to the sky" and enjoyed them, marking the tables of information on the planets with exclamation points and question marks. But I paid special attention to the gaps in the tables; where the word unknown appeared; where the authors had inserted question marks after certain entries. When we arrived at the planetarium that evening, I was prepared to listen and to learn.

We watched the planetarium show on the January sky with avid interest, and I stayed after the crowd had left to pepper the speaker with questions about the Moon and Mars. On the way out we passed by a meteorite display featuring a large pitted piece of metallic iron. I explored it animatedly, trying to imagine where and how it had formed. Suddenly a jolt like an electric shock ran up my spine. I stiffened and became utterly still.

My mother looked at me strangely. "What's the matter, Tommy, don't you feel well?"

I caught my parents by their arms and stared at the meteorite, gathering my thoughts. I slowly turned to look at them with great intensity.

"Someday—I'm going to be very, very rich." My parents glanced at each other in puzzlement.

Later, in the car on the way home to New Jersey, I drowsily reviewed the highlights of the evening. I remember I was humming a little tune to myself as I lay stretched out on the back seat.

"Dad?"

"Yes, Tommy?"

"There's something else I learned tonight."

"You mean, besides about getting rich?"

"Yes. Someday I am going to go where that meteorite came from."

But the strangest tale is yet to tell: I was right.

—From *How the Future Was Won*
The autobiography of Thomas L. Duncan
Kosmos Press, St. Petersburg
Special Sesquicentennial Edition (2091)

Western culture was born into a tiny universe, in which Earth was the center of creation. The Sun, the Moon, Mercury, Venus, Mars, Jupiter, and Saturn orbited about Earth—according to Ptolemy, writing in the second century A.D., in strange patterns of nested cycles and epicycles. Their motions, no matter how complex they appeared, had to be analyzed in terms of superposed circles, because circles are the purest and most perfect geometrical form. These seven celestial wanderers (*planets*, from the Greek word for "wanderers") were regarded as perfect, immaterial spheres, a complete and immutable set. No new bodies could be added to this list because seven is a mystical number, a sign of completeness and perfection. When Galileo, observing the heavens in 1610 with the newly invented telescope, claimed the discovery of four large moons in orbit about Jupiter, his conclusion was unacceptable. Prominent churchmen concluded that the telescope therefore had to be an unreliable instrument; in fact, it had to be a tool of deception sent by the Adversary to lead humans to question the church's authority. Evidence from telescopic observations therefore had to be deemed inadmissible. Indeed, since the church held that the heavens were perfect and divine in origin, any observation at odds with that theory, such as mountains on the Moon or spots on the Sun, had to be rejected.

Just beyond the realm of the Moon, Sun, and planets, the story went, there was a great black vault pierced by a myriad of tiny holes. Beyond that vault was the Celestial Kingdom, a world of pure light, whose presence was revealed to us only by those tiny holes in the vault of heaven. The entire physical universe in this conception was only a few times larger than the distance from Earth to the Moon. Light could cross this universe in a few seconds: in modern language, we would say that the universe is only a few light-seconds in diameter.

The planets, being immaterial perfect spheres, had to be featureless. Questions such as What made the craters on the surface of the Moon? How old are the volcanoes on Mars? What is the atmosphere of Venus made of? could not be answered because they could not even be asked. The modern reader must wonder how such transparent nonsense could survive a single glimpse of the complex, tortured surface of the Moon. But so entrenched was the equation of the physical heavens with the spiritual (perfect) heavenly realm that it was necessary to dismiss the features seen on the Moon as a simple reflection of the Earth on the Moon's perfect mirrorlike surface. Medieval texts even attempted to correlate the face of the Moon with the map of Earth. The dark areas of the Moon were taken to correspond with Earth's seas, a notion perpetuated to this day by the Latin name *maria* (MAH-ree-uh, meaning "seas") given to them.

The stars, being imagined as tiny holes in the dome of the firmament, all had to be at the same distance from Earth. Although no secure means to measure the distances of the planets and stars yet existed, it was widely accepted that the universe was a few hundred thousand to a few million kilometers in diameter. By the time the seventeenth-century Dutch physicist and astronomer Christiaan Huygens discovered a method of measuring the speed of light by careful timing of the eclipses of the satellites of Jupiter, he found that the known solar system spanned at least two billion kilometers, a distance that would take light about an hour to cross, not just a few seconds.

What then were people to make of fireballs, meteorites, comets, meteor showers, and other heavenly spectacles not accommodated by the theory? It was necessary to deny that they were planets, since the number of planets had to be seven. Meteorites could not have been of celestial origin, because the heavens were perfect and immaterial, certainly not the kind of place where ugly rocks might have been found. Meteorites therefore had to be atmospheric manifestations native to Earth. A popular theory of the eighteenth and nineteenth centuries held that they were made in the upper atmosphere by condensation of gaseous exudates from Earth. Because of the association of most meteorite falls with brilliant celestial fireworks displays and loud thunderlike rumblings, it seemed likely that these gaseous exudates somehow ignited and burned high in the atmosphere, giving rise to meteorites as the "ash" of the violent combustion process. Others held that meteorites were more immediately terrestrial in origin, being akin to volcanic bombs ejected by violent eruption events. But in the late eighteenth century the infant science of chemistry continued to cast doubt on these ideas, repeatedly finding meteorites with compositions that differed strikingly from all known types of terrestrial volcanic rock.

Comets presented a more serious problem: they could not be mere local atmospheric phenomena because they could be viewed from half the Earth at the same time. They in some sense had to be celestial, but they were also erratic and unpredictable: they appeared unforeseen, changed dramatically in position and appearance in a short time, and then vanished. The general opinion was that comets were immaterial, spiritual portents sent by the Creator as warnings about impending momentous events. A voluminous, highly spirited, and utterly contradictory interpretive literature grew up about comet apparitions. A common interpretation of the appearance of a new comet was that it was a signal of the impending birth or death of some great person or of an impending victory or defeat in battle. A comet appeared at the time of the death of Julius Caesar (44 B.C.). The appearance of a bright comet in 1066 is immortalized in the Bayeux Tapestry and connected by tradition firmly to the Norman Conquest. In his famous *Adoration of the Magi* (1303), Giotto reflected the medieval obsession by inserting a comet at the scene of the birth of Christ, drawing on the 1301 apparition of Halley's comet for his artistic inspiration. Deaths occurring months or even years after the appearance of a new comet were assigned to these "portents"; indeed, some great comet sightings were even linked to events that had occurred months or years *before* the comet appeared, making them very poor portents indeed. If no correlation with a great event was obvious, then that was simply a shortcoming on the part of human observers, as ignorant interpreters of the divine mind.

The tendency of meteor showers to occur in the same parts of the sky on the same day of different years suggested to some scientists that they were in fact following orbits about the Sun and were subject to prediction. Nonetheless, great feats of mental gymnastics were performed to make them into atmospheric phenomena. Their tendency to reappear at fixed times of the year was attributed to seasonal variations in the release of the putative "exudations" by the surface of Earth. The prominent series of meteor showers seen each August (when people were most likely to be outside at night) was cited to argue that the exudations were baked out of the ground by the heat of summer. The annual cycle of faint meteor showers in November, and the occasional apparitions of spectacular showers in that month at intervals of roughly thirty-three years, were sometimes admitted as embarrassments to the "summer exudation" theory.

As astronomers came to appreciate the differences between stars, planets, comets, and meteors, the true architecture of the heavens began to emerge. The stars were clearly more distant than the other bodies, but more precise understandings had to await the invention of techniques for measuring their distances.

A more modern awareness of the real relationship of Earth to the solar system, stars, and the universe first dawned in the public imagination in the early nineteenth century, on the heels of the American and French Revolutions. In 1838 the great German mathematician Friedrich Bessel reported the first crude measurement of the distance of one of the nearest stars, 61 Cygni. That distance was roughly six hundred thousand times the distance from Earth to the Sun. Many other distance measurements of relatively nearby stars soon followed. Viewed at such immense distances, even the Sun itself would have been only about as bright as those faint-looking stars in the night sky. Those stars must then in fact have been as bright as the Sun to be visible over such distances. The conclusion was astonishing: The universe must be so large that it would take light many years to cross all of known space. Each star in the sky was in fact a sun in its own right, possibly attended by its own family of planets. The laws of Newtonian physics and mathematics might then apply to, and help us to understand, a vastly larger universe than we hitherto thought possible. The church, both Catholic and Protestant, endorsed the authoritative opinion of Aristotle that the heavens and Earth were of different nature and obeyed entirely different laws. Isaac Newton himself, through his theory of universal gravitation, showed that, contrary to Aristotle's view, one law of gravitation applied to apples and the Moon alike. Some were exhilarated at the prospect of an immense universe whose properties, scope, and laws could be uncovered by the tools of science. Alexander Pope wrote:

Nature and Nature's law lay hid in night;
God said, "Let Newton be!", and all was light.

Others found the idea threatening. The seventeenth-century French savant Blaise Pascal recoiled from the prospect of a vastly larger universe, centered neither on Earth nor even on its Sun: "The eternal silence of those infinite spaces terrifies me!"

How Pascal would have recoiled from the modern measurements of the universe! What would he have said if he had known that the "vast" starry realm known to him was but a tiny corner of an immense "island universe" of stars, the Milky Way spiral galaxy, tens of thousands of light years across, containing several hundred billion star systems? And how would he have reacted to the knowledge that there are billions of other galaxies comparable to our own, spanning distances out to about sixteen billion light years from Earth? The volume of the universe known today is roughly 10^{27} (1,000,000,000,000,000,000,000,000,000) times as large as that which "terrified" Pascal!

What was the role of humankind in this vast new universe? The cozy dome of sky and crystal spheres of the Ptolemaic universe, all firmly centered on Earth, affirmed humanity as the central fact and focus of Creation. But Copernicus shattered the crystal spheres and put the planets in orbit about the Sun; then Bessel banished the stars to immense and "terrifying" distances. Discoveries of the twentieth century then exiled the Sun to the sparse outer fringes of the Milky Way, a rather typical spiral galaxy, just one among billions. As it became accepted that the bodies in the heavens were material objects, not spiritual beings, extension of Newton's laws of motion, first to the Moon and the planets and then to the stars, became possible. Physics gave us, in Bill Bradley's phrase, "a sense of where we are"—but also a sense that the laws governing *our* motions in space were the same ones as those governing the planets. Modern astronomy, while opening an immense gulf of space and time between us and the stars, had at the same time established a world in which we and the stars all played by the same rules. The gap between a spiritual heaven and a material Earth, formerly unbridgeable, had now shrunk to manageable proportions: it was now only a *physical* gulf that separated us from the Moon, the planets, and even the stars. True, that gulf was still very large. But the physical gulf between the continents had loomed equally large in the minds of fifteenth-century geographers and explorers. From Earth to the Moon was indeed farther, but to certain visionaries of nineteenth-century technology, the gap was *of the same kind* as that which separated the continents.

It was not until the general acceptance of the material nature of the heavens that it became possible to even conceive of space travel. Early literary portrayals of travel in space, such as Daniel Defoe's *The Consolidator* (1705), had insufficient contact with reality to give them any credence as predictions, being little more than fantastic settings for political and social satire.

The very idea of space as a hard (deep) vacuum was unfamiliar to some early writers, and as recently as the early 1900s it was common for the means of propulsion used to reach other worlds to be either physically fanciful ("lifted by the Sun's attractive force upon vials of dew") or utterly nonphysical ("transported to Mars as in a dream"). It was with the birth of modern science, engineering, and science fiction that the actual *means* of travel and survival in space became accessible to rational discussion. Jules Verne envisioned, in *From the Earth to the Moon* (1865), a giant cannon built in Florida (and based on Civil War military technology) to propel a sealed "spacecraft" to the Moon. We know today that the firing of this gigantic gun would have utterly crushed its delicate human occupants, but at least Verne attempted to propose a specific physical method to fire vehicles into space. He imaginatively conceived of a moment of weightlessness experienced by the astronauts as they pass through the equal-gravity point between Earth and Moon (actually, once free of the retarding frictional effects of the atmosphere, occupants of the projectile would remain weightless right up to the moment of impact with the lunar surface). Not long thereafter, H. G. Wells postulated physical means of travel between the planets in his *War of the Worlds* (1898).

The idea that technology could actually make the heavens accessible to humans dates only from the beginning of this century and the pioneering (and independent) work on rocket propulsion by Konstantin Tsiolkovskii in Tsarist Russia, Robert Goddard in the United States, and Hermann Oberth in Germany.

Konstantin Eduardovich Tsiolkovskii was born in 1857 in the village of Izhevskoye in Russia's Ryazan Province. A severe bout of scarlet fever as a boy left him so deaf that he was unable to attend school. Home-schooled by his parents until age sixteen, Tsiolkovskii read voraciously and developed a lifelong fascination with mathematics and physics. Having exhausted the educational opportunities of his village, he was sent by his parents to Moscow to study chemistry, astronomy, mathematics, and physics. There his studies led him to the reading room of the Chertkov Library, where he fell profoundly under the influence of the philosopher Nikolai Fyodorov, a cataloguer at the Chertkov. Fyodorov was the proponent of an astonishing philosophy that envisioned the perfection of mankind, the raising of the dead, and the settlement of the cosmos.

When young Tsiolkovskii's health deteriorated after three years in the harsh Moscow climate, he returned home to become a teacher of mathematics and physics in the village of Borovsk. At age 24, he wrote several original papers on gas kinetic theory, which he sent off to the St. Petersburg Society of Physics and Chemistry. The eminent chemist Dmitri Mendeleev responded personally, informing Tsiolkovskii that the ideas he set forward were valid, but that they had been published in England some twenty years earlier. Sensing

Tsiolkovskii's intellectual powers, he nonetheless strongly encouraged Tsiolkovskii to continue with his research interests.

Working alone in the early 1880s, while employed as a schoolteacher in Kaluga, he pioneered the theory and experimental testing of lighter-than-air craft. His interests then turned toward the engineering and physiological problems of space flight. He explored the mathematics and physics of rocket propulsion at a time when no one had the means to carry out experiments to test his ideas. His greatest work, published first in 1903, was *The Exploration of Cosmic Space by Means of Reaction Motors*, the world's first monograph on rocket propulsion. In it he developed the mathematics of rocketry and introduced the idea of the "step-rocket," in which several stages are fired sequentially to achieve orbital velocity and even escape velocity from Earth. Until the 1920s his work remained utterly unknown outside Russia.

Experimental work on rocket propulsion did not begin in Russia until long after the 1917 revolution, when Tsiolkovskii was already an old man. The GIRD Hydrodynamic Research organization, a private group dedicated to testing and developing flight-rocket hardware, began operation in the early 1930s under the leadership of M. K. Tikhonravov and Sergei P. Korolev. The latter was to achieve lasting fame as the "chief designer" of the Soviet space program in the 1950s and 1960s. Tsiolkovskii himself never received a kopek of government funding until he received a pension.

Tsiolkovskii envisioned the conquest of space in fourteen points. The first few concerned elements of basic rocket technology of little interest to us here. Point 5, unmanned orbital flight, was to be followed by point 6, manned orbital flight, and then by the development of the space suit and its use in space walks (points 7 and 8). Next was to come the development of space agriculture (9) for use in orbiting space colonies (10) and the use of solar power for transportation and power in space (11). Once the technology to operate in space was in hand, humankind would turn to the exploitation of the asteroids (12) and use them as the basis for the industrialization of space (13). Tsiolkovskii's fourteenth and final goal, echoing Fyodorov's visionary philosophy, was the perfection of humanity and society.

Robert Hutchings Goddard, born in 1882 in Worcester, Massachusetts, first developed an interest in rocketry and space travel as an undergraduate at Worcester Polytechnic Institute. After receiving a Ph.D. from Clark University in 1911, he stayed on as an assistant professor to teach and to pursue largely theoretical research on rocketry. By 1909 he was already aware of the superior potential of liquid hydrogen–liquid oxygen rocket engines. In a manuscript written by Goddard in 1918, and kept sealed at his insistence for fifty-four years, he envisioned the expansion of humans into space through the use of space materials, culminating in the construction of a vast fleet of nuclear-powered asteroids that would carry mankind to safety from the death throes of the Sun countless years in the future. In 1919 his vastly more conservative monograph, *A Method of Reaching Extreme Altitudes*, was published by the Smithsonian Institution. It was variously regarded by its readers as brilliant or deluded. By 1920 Goddard, like Tsiolkovskii before him, had turned to the theory of liquid-fueled rockets as the most promising means for launching Earth satellites. In the 1920s he struggled to raise funds for the actual construction and testing of small liquid-fuel rockets, amidst an atmosphere of mixed hostility and skepticism. Nevertheless, the Guggenheim Foundation came forward with the funds necessary for Goddard to begin flight testing in New Mexico. Even as his work bore fruits with successful test flights of his first small rocket in 1926, Goddard was pilloried as a "Moon-mad" crackpot in an infamous editorial in *The New York Times*. Goddard's research on the mathematical and physical basis of rocket propulsion clearly demonstrated that the performance of rocket engines improves significantly when they are fired in the vacuum of space. He also developed, quite independently of the earlier but inaccessible work of Tsiolkovskii, the theory of multistage rockets. It is interesting that Goddard's achievements were not in any way dependent on government foresight and vision. It was the Guggenheim Foundation, not a government bureaucracy, that saw the promise of Goddard's work and invested in it. The first government "support" for the work of Goddard was awarded in 1960, fifteen years after his death, when the U.S. government was found guilty of infringing his basic patents and was fined $1 million.

∴ ✻

Hermann Julius Oberth, born in 1894 in Hermannstadt, Transylvania (then a part of the Austro-Hungarian Empire), studied medicine, mathematics, and physics in Munich until being called into military service during World War I. In 1915 he presented the design of a long-range liquid-fuel rocket to the Austro-Hungarian War Ministry, which rejected his idea as a "fantasy." After the war, Oberth moved to Heidelberg to study with the outstanding chemists and physicists there. In his doctoral dissertation, completed in 1922, he developed the mathematics of space flight and demonstrated how a rocket could be fired to reach escape velocity from Earth. In that same year he heard for the first time of the 1919 monograph by Goddard, with whom he began corresponding on matters of mutual interest. His brilliant doctoral dissertation was rejected by the University of Heidelberg, but Oberth was so convinced of its worth that he had it published in 1923 at his own expense. In 1925 this publication brought to him correspondence that informed him of the existence and work of Tsiolkovskii, and they began to correspond. Oberth's book *Wege zur Raumschiffahrt* (Routes to Space Travel), published in 1929, won the international Hirsch Prize of ten thousand francs, which Oberth immediately applied to experimental research on rocket propulsion. That book contained a proposal for electrical propulsion of spacecraft, a technique we would now call ion propulsion. The publicity surrounding this book also stimulated the newly formed (1927) Verein für Raumschiffahrt (VfR), literally, the League for Spaceship Travel. The success of this organization in firing small liquid rockets in turn attracted the attention of the German Army (Bundeswehr) well before the outbreak of the World War II. The Bundeswehr was receptive not so much because of the obvious merits of rocketry but because of the Treaty of Versailles, which placed very severe restrictions on the development of German artillery but failed to mention rocketry.

Following Goddard's first flight test of a liquid-propellant rocket in 1926, events accelerated markedly. Stimulated by Goddard's success, the American Interplanetary Society was founded in 1930. The AIS was later to become the American Rocket Society and function more as a trade organization than as a research and education foundation. The vigorous British Interplanetary Society (BIS) was founded in 1933. Thus, by 1933 the four leading technological powers of the world had all established nongovernmental organizations of enthusiasts dedicated to building and testing rocket hardware and exchanging ideas about humankind's future in space.

Progress in rocket performance closely followed establishment of these organizations. In 1931, Goddard held the world altitude record for a rocket, with a flight to 1,700 feet. By 1933 the Soviet GIRD group had flown a rocket to 1,300 feet, and in 1934 the ARS fired a rocket over a horizontal distance of 1,338 feet. With Wehrmacht funding of eighty thousand marks, a German team fired a small rocket, the A2, to a height of 1.5 miles in 1934. In 1935, Goddard fired two rockets to ranges of 1.8 and 2.6 miles. The German success stimulated the infusion of eleven million marks from the Luftwaffe and Wehrmacht in 1935. In 1937 a large new test range was opened on the island of Peenemünde, on the Baltic coast. By 1941, Oberth was at Peenemünde.

Development of military rockets was proposed to Hitler in 1939 and rejected by him on the basis of a bad dream in which a rocket failed to work. Working on a shoestring budget, the Bundeswehr rocket team carried out a successful test flight of the A4 rocket in October 1942. A movie of the test flight was immediately sent to the Führer, who refused to watch the film for a full year. But when he finally saw the film, in 1943, Hitler had a change of mind and authorized greatly increased funding for development of a ballistic missile capable of bombarding London from the Continent. The production model of the A4, renamed Vengeance Weapon 2 and abbreviated V2, appeared in service in September 1944. By then D day was history, and the Allies had recaptured Paris. The first V2 launches were directed against both London and Paris, rather than being concentrated on England.

The V2 missile lacked an electronic guidance system and was woefully inaccurate, generally landing ten or more kilometers from its target point. This meant that the V2 was ineffectual against military point targets but could be fired effectively against extended target areas twenty or more kilometers in diameter. The only such targets available to the Nazis were urban areas. Thus, the V2 was incapable of serving as a counterforce weapon, but it made a perfect terror weapon against civilian populations—and that is precisely how it was used. Fortunately, casualties from these attacks were light. The 4,300 V2 rockets fired against England caused only 2,500 deaths, almost all of them civilians.

The large-scale military use of the V2 in the closing days of World War II demonstrated the reality of rocket propulsion. The dismantling of the German rocket program at the end of the war left the technical masterminds of the program and most of the surviving missiles in American hands, although most of the factory personnel and equipment fell to the Russians. Both sides assimilated what they could of German rocket expertise, but it was the Americans who got Wernher von Braun and the rest of the design team, the true brains of the operation. A number of captured and modified V2s were launched at White Sands, New Mexico, and von Braun's team went to work at the United States Army's Redstone Arsenal (in Huntsville, Alabama) designing short- to medium-range (10 to 500 miles) tactical battlefield missiles.

Opinion remained divided on the possible utility of intermediate- and intercontinental-range (1,500 miles and 5,000 miles) weapons. As late as 1947, no less an authority than America's wartime technical genius Vannevar Bush, the head of the U.S. Office of Scientific Research and Development, interpreted the performance of the V2s over London as evidence that long-range missiles could not possibly hit point targets and hence were of no military significance. Other observers reached the more practical conclusion that some means of controlling the flight of rockets were needed. They therefore advocated the development of electronic guidance systems for military missiles, beginning with vacuum-tube technology. Yet others of a more visionary turn foresaw the evolution of the rocket into a device to deliver payloads into orbit about Earth, or to send them to the Moon and the other planets. The German engineer Willy Ley wrote several books popularizing the idea of expeditions to the Moon and Mars, and he lit a spark in the public imagination.

In 1947 the Soviet Union initiated a program to design and build intercontinental-range ballistic missiles (ICBMs). The postwar American ICBM program was canceled by President Harry Truman, at the urging of Vannevar Bush, in the same year. For six fateful years Stalin backed the Soviet ICBM program while the United States slept. Only the decision of Convair/General Dynamics to continue ICBM development on a shoestring with its own funds prevented the Soviets from achieving an insurmountable lead. Finally, in 1953, American government funding for ICBM development was restored by the Eisenhower administration. But by then the Soviet program had a hefty lead that guaranteed them first entry into space, with the launching of *Sputnik 1* in 1957.

As the readers (and writers) of newspapers became aware of the rapid progress of rocket technology, editorial writers rather abruptly ceased writing curmudgeonly polemics ridiculing space and its advocates, and commenced writing glowing Sunday-supplement articles, lavishly illustrated, extolling the promise of space and the virtues of its practitioners. It also suddenly became possible to say kind words about space visionaries. For example, Arthur C. Clarke, a successful British engineer turned science fiction writer, had proposed in 1947 that satellites could be placed in orbit forty thousand kilometers above the equator, where their orbital periods of twenty-four hours would cause them to remain stationary in the skies as seen from the ground. These satellites, Clarke suggested, might be used as communications relays to "bounce" radio and television transmissions across the oceans and enable global telephone and telegraph service without reliance on transoceanic cables. With the launching of the first Earth satellites, Clarke underwent a near-instantaneous transformation in the media from harmless crank to brilliant prophet. Another modest proposal of Clarke's, published in 1939 in the *Journal of the British Interplanetary Society*, suggested that missions to the Mars system might be able to extract water and other volatile materials from the Martian moons Phobos and Deimos. These resources would then be processed to make air and water for the crew and rocket propellants for use in landing on Mars, taking off from Mars, and returning to Earth. The press generally ignored this suggestion. Perhaps it was simply too "far out" to be worthy of repetition.

Other visionaries fared similarly. The British physicist J. D. Bernal wrote in his famous essay "The World, the Flesh, and the Devil" (1929) of the engineering of asteroids into vast hollow habitats to hold immense populations, carry them throughout the solar system, and eventually ferry them to other stars. He argued strongly that the cost of launching materials out of Earth's deep gravity well was prohibitive compared to using the resources already available in space. He was the first to propose that solar power could be collected in space and beamed down to Earth to power our industries. He envisioned vast, hollow space habitats built of asteroidal materials and powered by the Sun, each engaged in the task of replicating itself and multiplying until the competition for sunlight became so severe that they would embark on interstellar voyages to find new resources for growth. The extraordinary mind of the British science-fiction writer Olaf Stapledon surpassed even this astonishing vision. In *Last and First Men* (1930) he foresaw a vast panorama of future proliferation of mankind in space, giving rise by means of directed evolution to new species adapted to conditions on other planets. Space, Stapledon implied, was not just a list of dangers; it was a menu of opportunities.

The reader may have noticed a peculiar coincidence: all three of these visionaries were British. It was their brilliance and farsightedness that stimulated the formation of the extraordinary BIS. This society, through its meetings and publications, has for over sixty years consistently advanced the cause of innovation and exploration in space. Its membership, an eclectic mix of people with a deep interest in space, spans the political and social spectrum and touches many nations. Yet, strangely, the BIS has had less influence on British space policy than its counterparts in the United States, the Soviet Union, and Germany. Perhaps the BIS members were so farsighted that they lost touch with the mundane necessities of day-to-day political life. Certainly there has been little inspired and visionary support of space activities by the British government. Whatever the reason, it remains true that the greatest triumphs and aspirations of the space-faring nations represent to a remarkable degree the practical working out of British dreams.

One might suppose that the high-water mark of space exploration to date, the Apollo program for exploration of the Moon, was in part inspired and motivated by the ideal of self-sufficiency in, and profit to mankind from, the material and energy resources of space. But this was not the case. The goal of Apollo was not practical benefit, nor even scientific research, but political posturing. In the Apollo exploration of the Moon, the concept of using lunar materials was intentionally avoided because of engineering conservatism and managerial short-sightedness. The idea was not to build a foundation for the future but to get there quickly, boast about beating the Russians there, and then cancel the program to "cut the losses." The capability to continue to explore space using the Apollo spacecraft and the Saturn boosters was swiftly destroyed once the political goal had been met, preventing the space endeavor from evolving into some mere commercial (profitable) or curiosity-driven (scientific) venture that might take on a life of its own and escape governmental control. It is a measure of our time that the phrases *commercial* and *curiosity-driven* have again become pejorative, almost as damning as *visionary*. Apollo itself, widely touted by the press as visionary, was in essence a political act intended to make America look good compared to the Soviets—a particularly strongly focused reflection of the cold war. Our inheritance from Apollo was not the realization of a dream, but rather its end. But as long as human beings live, they will feel curiosity and engage in commerce. The only viable route to a future of growth is to allow these basic human activities free rein.

What do we know, then, about the resources of space and how to use them? The scientific study of the samples returned from the Moon by the Apollo expeditions, and data sent back by a variety of American and Soviet unmanned missions, have given us a vast library of information about the availability of resources on the Moon. This will serve as an excellent place to start our survey of the resources of the solar system, the first halting step in a journey of exploration that will eventually take us very far indeed in both space and time.

3

A POCKET GUIDE
TO THE MOON

I read everything in the newspapers and magazines about rockets (then mostly the V2) and astronomy. A year or two after it came out, I read John W. Campbell's novella The Moon Is Hell! *which dealt utterly unbelievably with attempts of a small lunar base to achieve self-sufficiency. The book is, in general, pretty awful, but the challenge of survival on the Moon intrigued me. And of course, the chief protagonist of the story is one "Thomas R. Duncan, Ph.D."—enough to guarantee my attention! In the eighth grade, in the spring of 1954, I gave a talk to my class about rockets and artificial satellites. They all laughed. In my high school years, I made several trips to the Rutgers and Princeton libraries and followed ideas back to their roots: a young engineer named Arthur Clarke writing in the* Journal of the British Interplanetary Society *in the 1930s about making rocket propellants out of the moons of Mars; Olaf Stapledon's mind-bending tales of the distant future of mankind, with our genetically tailored descendants living among the planets and then among the stars; J. D. Bernal's essay "The World, the Flesh, and the Devil," pointing the way to the use of the material and energy resources of space to meet the coming needs of Earth. I read brief accounts of Konstantin Tsiolkovskii's and Robert Goddard's pioneering visions of the development of the rocket as a key to the opening of the solar system. I read Doc Smith's stories about men and women who flitted about space fighting galactic conspiracies and righting cosmic wrongs—but when all the pulpy plots, the swashing and buckling, were done, what stayed with me was his image of Kim Kinnison in* Gray Lensman, *posing as a crusty old asteroid miner, hauling in a big piece of platinum. I always pictured it as looking exactly like that meteorite at the Hayden.*

I graduated from Passaic High School in 1958, just a few months after Sputnik 1, *and enrolled at Princeton to study aerospace engineering. In 1961 I heard Luigi Crocco's lecture proposing a grand tour of the Jovian planets. I took enough physics courses to please, and enough astronomy courses to upset, my adviser in aero. Soon thereafter, while a graduate student in La Jolla, I read Poul Anderson's "The Rogue" and "Say It with Flowers," later to become parts of his* Tales of the Flying Mountains. *The great Nobel laureate Harold Urey was there then, and I went to his lectures on the Ranger missions to the Moon. Urey emphasized the importance of the chemistry of the Moon as a clue to its origin. I also sat in on an undergraduate course on the solar system, the exact sort of course I had been searching for since I was six or seven.*

In my world, it didn't matter whether the source of an idea was science or fiction. If it made sense, or if it could be made to make sense, then it was mine. I was not uncritical, but I was very eclectic and very persistent. I read dozens of books on strange powers (my own experience had shaken me profoundly) but found them full of ignorance and superstition, written by Believers, not by those with a true personal experience to relate. I read all the flying-saucer books, and threw out 99 percent of their contents as trash, lies, delusions, and baseless speculation. Then I found Charles Fort's catalogs of bizarre events. I howled with laughter at his mock-theories, but sifted his vast body of evidence with great care, looking up hundreds of his references to the source literature. I had taken bits of French, Latin, German, and Russian, and I had listened to enough operas to pick up an idiosyncratic dash of Italian and Spanish. I used these languages to probe the frontiers of the possible, digging through vast amounts of source materials. And I found those frontiers. Oh my, yes! Amid the tons of refuse there were a few precious gems. The first jewel in the crown was the realization that the most promising frontier of all was the one where science found the paths and engineers built roads. And the centerpiece of the diadem was the knowledge, *the absolute unshakeable* perception, *that the Moon and planets were not just little flecks of light on the dome of the sky; they were* other places, *and those places were* not *just like home.*

—From *How the Future Was Won*
The autobiography of Thomas L. Duncan
Kosmos Press, St. Petersburg
Special Sesquicentennial Edition (2091)

∴ ✱

Bridging the void between Earth and our Moon, arguably the greatest single feat of human technology, has brought us face to face with a new world. Step by step, first through astronomical observations and later through the eyes of automated spacecraft and human explorers, the true magnificence and utterly alien integrity and beauty of the Moon have emerged. Long before the natives of Earth learned how to build rockets and spacecraft, astronomers had discovered that the Moon is a barren body devoid of atmosphere and liquid water. The "magnificent desolation" of the lunar surface makes us wonder whether it would even be possible for people to live there, let alone thrive. The lunar surface, baked in daytime by the scorching heat of the naked Sun and chilled at night by exposure to black space, is a desert to a degree unachievable on Earth. But we have made deserts bloom before: could we do the same thing on the Moon?

In the pre-space age era, astronomers measured the size and shape of the Moon and analyzed its motions with precision. The near side of the Moon was thoroughly photographed and measured. Nearly every feature larger than a kilometer on the entire near side of the Moon was mapped, and the larger ones were given names.

The Moon does not rotate with respect to Earth. Further, the Moon is not spherical: all three of its principal diameters differ by several kilometers. The diameter along the line pointing toward Earth is the largest, the diameter along the direction of its orbital motion is intermediate, and the polar diameter is the smallest. The elongation of the Moon along the axis that points at Earth constitutes a permanent, "frozen" tidal bulge that is locked onto Earth's gravity. The portion of the

Moon that points permanently toward Earth, the near side, is of course the part that is most easily observed from Earth.

Most of the near side of the Moon is bright, rough, high terrain, called the lunar highlands. Several large dark splotches, the mare (MAH-ray) basins, cover about a quarter of the visible hemisphere. Unlike on Earth, most of the visible features on the Moon are circles or great arcs constituting parts of circles. The highlands, seen through a large telescope with high magnification, break down into a vast and complex pattern of superimposed circles. Craters (from the Greek word for "cup") are everywhere, small craters within large ones, layer upon layer, generation upon generation, overlapping and obliterating one another. The craters are named after great scientists of the past: Tycho and Copernicus; Aristarchus and Kepler; Messier and Eratosthenes.

Most of the mare basins also have circular outlines but have few craters within them, showing that they have not been exposed to the rain of meteorites for as long as the cratered highlands. The mare basins have dark, rather smooth floors that lie well below the altitude of the highlands. The great circular basins of Mare Serenitatis (Sea of Serenity), Mare Tranquillitatis (Sea of Tranquility), and Mare Crisium (Sea of Crises) sport circular rims of mountains, suggesting that they too are just oversized craters. Some areas of the dark, smooth lowlands are, however, extremely irregular in shape. The largest, Oceanus Procellarum (Ocean of Rains), sprawls over a large area without the least suggestion of circular outline or crater walls.

Folded mountain chains, which are common on Earth, are utterly absent on the Moon. The horizontal forces that cause the formation of these structures on Earth are the result of continental drift, which is absent on the Moon.

Many of the larger craters in both the highlands and mare basins display clusters or rings of steep mountains at their centers. There is a continuous sequence from small bowl-shaped craters to larger craters with simple central peaks to even larger craters with central mountain rings. The largest craters show, in addition to their rim walls and central peak structures, outer concentric rings of mountains. One feature partially visible on the eastern edge of the Moon as seen from Earth is the Mare Orientale (Eastern Sea), a large multiringed basin with a dark, smooth floor.

Despite the prevalence of "seas" and "oceans" on the near side of the Moon, telescopic inspection reveals absolutely no evidence of the presence or former presence of water anywhere on its surface. There are certainly no seas and oceans. Despite features with names such as the Marsh of Decay (Palus Putredinis) and Rainbow Bay (Sinus Iridum), there are no marshes, gulfs, bays, channels, rivers, lakes, sloughs, swamps, creeks, or bights. There is no sign of any of the effects of running water; no ancient shorelines or beds of sediment, no abandoned stream channels, and no deltas. Such aridity is surely no recipe for an Eden in the sky. But is there water there that does not so readily meet the eye?

The most sensitive searches with spectrometers for traces of water vapor and other gases show clearly that the Moon's surface is a very hard vacuum, emptier than the best vacuum attainable in laboratories on Earth. Water vapor is so rare that liquid water cannot possibly be present on or near the lunar surface: if there were liquid water, then it would boil vigorously in the lunar vacuum, producing vast clouds of water vapor. Its evaporation would supply an amount of water vapor billions of times larger than the limit set by spectroscopists. The observations do not, however, rule out the presence of extremely cold ice, which would have a very low vapor pressure. The only place on the Moon cold enough for ice to persist would be in permanently shadowed crater bottoms very close to the lunar poles.

Even before the first spacecraft missions to the Moon, the dark materials that cover the floors of large craters and mare basins were suspected to be lava flows. Numerous faint, sinuous wrinkle-like ridges and rilles seen by telescopic observers on the dark mare floors were attributed to repeated flooding by dark, highly mobile lavas. The ideal material seemed to be basalt, a dark volcanic eruptive rock, rich in iron oxides, which forms a highly fluid melt. On Earth, basalts are common volcanic rocks that form extensive lava flows, such as in the Columbia plateau of Oregon, Washington, and western Idaho and the Deccan plateau in India.

The discovery of lava flows that cover vast distances on the Moon provided an interesting clue to their composition. Viscosity, or resistance to flow, is a property of fluids containing long molecular chains that tangle and intertwine. Syrupy liquids such as honey contain bulky sugar molecules that bump into and stick to each other, raising the viscosity far above that of pure water. Solutions of unusually long molecules, such as proteins, often develop so high a viscosity that they form ropy or glassy liquids, of which Jello and consommé are familiar examples. In molten rock (magma) the main ingredients are oxides such as silica (silicon dioxide), lime (calcium oxide), magnesia (magnesium oxide), alumina (aluminum oxide), soda (sodium oxide), and various oxides of iron. Common natural rocks almost always contain abundant silica, and the minerals in them are mostly silicates of magnesium, iron, calcium, aluminum, titanium, and so on. (It is easy to keep these names straight with the following rule: If the name ends in -*ium*, it is a metallic element; if the name ends in -*a*, it is the oxide of that metal.) Every element has an internationally accepted one- or two-letter symbol, such as Si for silicon, Mg for magnesium, Al for aluminum, and so on.

In nature, the elements form stable chemical compounds with each other, usually involving oxygen. Naturally occurring crystalline solids are called minerals. Examples of important minerals on the Moon include two, olivine and pyroxene, that are composed of oxides of magnesium, iron, and silicon. Albite feldspar is also common. Many of the roughly three thousand known terrestrial minerals are chemically complex, and many require the presence of liquid water and gaseous oxygen to make them and keep them stable. By comparison, the few dozen minerals found on the Moon are usually quite simple in composition. In general, the main component minerals of lunar basalts are silicates of aluminum, magnesium, iron (Fe), calcium (Ca), sodium (Na), potassium (K), and titanium (Ti). Such rocks on Earth would not normally be mined for their mineral content.

The geological history of Earth has been long and complex, whereas that of the Moon was short and simple. The Moon's internal activity died while Earth was still in its geological childhood. Earth's tortuous history has caused repeated segregation and concentration of most of the elements into mineral deposits that occur, for each element, in at least a few places widely but inequitably scattered over the face of the planet. Ores, which are economically workable mineral deposits, are highly concentrated occurrences of useful minerals. Many ores are mined intensively, and several, such as iron oxides and the aluminum ore bauxite, play major roles in the terrestrial economy.

The most important factors in the formation of ore deposits on Earth are the action of liquid water, oxidation by the atmosphere, and the constant vigorous recycling and reprocessing of the crust. Whether comparably effective concentration processes have been at work to make rich, diverse ore bodies on the very dry and airless Moon is certainly debatable. Until evidence of such ore bodies can be produced, skepticism regarding their existence is fully justified.

Although large areas of the lunar near side's surface have been flooded with basalt, the mountains of the lunar highlands do not seem to be volcanic in origin. Instead, the basalt flows and volcanic central peaks are found to be very closely associated with large craters. Where there is evidence for volcanic activity on the Moon, it generally seems to be associated with impact events.

Long before spacecraft missions first explored the Moon, speculation on the origin of the lunar craters abounded. Some assumed that all the craters were of volcanic origin. Some attributed each crater to a bursting giant bubble of gases escaping from a vast ocean of magma. Yet others proposed that the craters were made by ancient impacts, possibly during the tail end of the era of accumulation of the planets from small bodies, several billion years ago. The resolution of this and many other questions about the origin, evolution, composition, and structure of the Moon had to await the era of spacecraft exploration.

∴∗∴

The first spacecraft missions to the Moon were launched surprisingly soon after the dawn of the space age. The Soviets, aware that their huge new ICBM gave them a several-year lead over the Americans, resolved to exploit their great weight-lifting superiority by pioneering a wide variety of new missions. Immediately after the successful launching of *Sputnik 1*, the chief designer, Sergei P. Korolev, received approval from Communist Party Chairman Nikita Khrushchev for a wide range of space spectaculars, including launching spacecraft capable of carrying a single cosmonaut (the Soviet term for astronaut).

Spacecraft and rocket upper stages for unmanned missions to the Moon, Mars, and Venus were all initiated.

The American response to the Soviet thrust into space was at first perilously close to panic. The press and many politicians clamored for a crash space program to assert American technical superiority. Khrushchev rattled his sabers vigorously, boasting (truthfully) that the launching of *Sputnik 1* and the much heavier, dog-carrying *Sputnik 2* demonstrated the existence of a very large intercontinental strategic missile, which we now know was called the R-7 by the Soviets. He also boasted that hundreds of Soviet ICBMs equipped with huge thermonuclear warheads were poised for instant launch against America in the event of trouble.

President Eisenhower, the leading American military authority of the time, strongly doubted the reality of the Soviet ICBM threat. He initiated the development of space-based photographic surveillance satellites and the Agena rocket stage to insert them into orbit. These spy satellites were to be in place within a few years to monitor all Soviet military activities. But there was an immediate problem that could not wait several years for resolution: did the Soviets in fact have hundreds of heavy ICBMs poised for instant launch at America? Eisenhower authorized overflights of the suspected Soviet missile bases by the high-flying U-2 spy plane with two history-making consequences: first, it was found that Eisenhower's skepticism was fully justified—there were no more than four operational ICBMs, and these took hours of preparation before they could be launched. Second, in 1960 a U-2 spy plane piloted by Francis Gary Powers suffered an engine problem near the central Siberian city of Sverdlovsk and descended to low enough an altitude to be shot down by a Soviet anti-aircraft missile. This event proved a great embarrassment to Eisenhower, even though the overflights had by then fully vindicated his position that there was no Soviet missile threat.

The first American response to the Sputnik program was a desperate attempt to launch a tiny "grapefruit" satellite with the Vanguard booster, which was still under development. In contrast to Soviet secrecy, the launch—which failed—was broadcast on live television, bringing further scorn to the American space effort. In further reaction to the launchings of *Sputnik 1* and *2* in 1957 and the three thousand-pound *Sputnik 3* in 1958, the United States, still without an operational ICBM, embarked on an ill-conceived crash program to throw together and launch a series of tiny lunar probes carried by small boosters. The first four Pioneer lunar probes, launched in 1958 and 1959 by the Thor Able, a booster based on the air force's Thor intermediate-range ballistic missile (IRBM), and the Juno II, based on the army's Jupiter IRBM, either blew up in flight or missed the Moon by heroic distances. Each launch was watched, photographed, televised, and followed to its sad denouement by the world press. The Moon seemed, at such times, to be hopelessly beyond reach.

In the midst of this circus, the Soviet Union announced the launch of its first lunar probe mission. The *Luna 1* spacecraft flew by the Moon at a distance of six thousand kilometers in January 1959. In September a second Luna spacecraft impacted on the near side of the Moon, the first human artifact to reach our natural satellite. A month later, precisely on the second anniversary of the launching of *Sputnik 1*, *Luna 3* was launched into a trajectory that intentionally missed the Moon by a narrow margin. After passing the Moon, *Luna 3* looked back, photographed most of the far side of the Moon, which had never before been seen, and televised several crude pictures back to Earth. These blurry, overexposed images revealed at least two small dark basins on the far side, embedded in vast expanses of bright highlands. These dark areas were named the Sea of Moscow (Mare Moscoviense) and Tsiolkovskii.

In hasty response, the new American Atlas ICBM, less than half the size of the Soviet R-7, was pressed into service. Equipped, like the Thor Able, with the upper-stage stack from the Vanguard rocket, the new Atlas Able was assigned the task of launching three more Pioneer lunar probes in 1959 and 1960. All three launches failed. The total American return to date from its lunar exploration program: seven launches, no successes, and no data about the Moon.

At that time the Atlas ICBM was still in flight testing. John F. Kennedy ran a vigorous and successful campaign for the presidency on the strength of the alleged "missile gap," not only accepting Khrushchev's wildly bloated claims at face value but also claiming that the Eisenhower administration had allowed the United States to slip perilously behind in strategic weapons technology. This attack was scarcely credible to the few who remembered clearly that it had been Truman who had canceled the American ICBM program in 1947 and the newly elected Eisenhower who had restored funding for the program in 1953. Truman's attitude toward space is well reflected by his widely quoted pronouncement that "this business of launching Earth satellites is a lot of hooey." Nonetheless, the fictitious missile gap carried Kennedy to the presidency in 1960 with a reputation as a visionary, dynamic young leader with an interest in space.

Meanwhile, the Atlas was overcoming its growing pains. The Agena upper stage, developed at Eisenhower's instruction to lift photographic reconnaissance satellites into space, was mated with the Atlas to make the new Atlas Agena vehicle. The Ranger program, dedicated to impacting the Moon at high speed with spacecraft carrying high-quality television imaging systems, was ready for its first flight on the Atlas Agena by the summer of 1961. But *Ranger 1* got stuck in low orbit around Earth, as did *Ranger 2* three months later. *Ranger 3* missed the Moon by a huge distance, and *Ranger 4* impacted on the near side of the Moon exactly on course—but its experiment package failed to work, and no photos were returned. In late 1962, *Ranger 5* approached the Moon—and missed it. *Ranger 6* again successfully impacted on the Moon, but again the TV equipment failed to function. The score as of 1963: thirteen American missions, no successes, and no data. Of the production run of nine, only three Ranger spacecraft remained.

Paradoxically, it was at this time of deep frustration that the enormous superiority of American space hardware first became evident. Using highly advanced solid-state electronics, solar cells, and radio and imaging hardware, American spacecraft in orbit about Earth could often outperform Soviet spacecraft that were 10 to 100 times as large. By having the head start into space, the Soviets led the way in debugging their equipment and getting it to operate for a few days in space. But that equipment was vastly inferior in design and in potential performance to American developmental hardware. All the Americans lacked was the experience to get it working.

It was in this setting that President Kennedy challenged the Soviet Union to a race to send astronauts to the Moon, a move intended to give free rein to advanced American technology and give Americans something to be proud of. NASA advised the president that a manned lunar landing was the first major goal the United States could expect to achieve before the Soviet Union. The Soviets entered the race with enthusiasm, but they did not publicly acknowledge doing so for fear that they might not win.

Meanwhile, the Soviets had launched two Luna-type spacecraft at the Moon in early 1963. One was stuck in low orbit about Earth, and it was not even acknowledged as a lunar-mission attempt until twenty-five years later. The other, apparently identical, spacecraft was intended to land on the Moon and survive, but the vehicle went astray and missed the Moon altogether. *Rangers 7, 8,* and *9,* in 1964 and 1965, were outstanding successes. Between them, they returned over seventeen thousand sharp pictures of the Moon, some taken so close that they showed details as small as a meter in size. By comparison, the *Luna 3* pictures of the far side of the Moon failed to show features as large as ten thousand meters.

The Soviet Union, in preparation for manned landings, then concentrated on landing automatic unmanned spacecraft on the lunar surface. Of five landing missions launched in 1965, all five failed. *Lunas 5, 7,* and *8* apparently performed as programmed, but they crashed into the lunar surface at high speeds and were destroyed because the "lumpiness" of the lunar gravity field was not yet understood. Finally, in January 1966, the *Luna 9* spacecraft carried out the first successful soft (survivable) landing on another body in the solar system. Thus, another key landmark was achieved by the Russians, an absolutely necessary step on the way to manned landings on the Moon.

Both the Soviet and American lunar programs concentrated on automated landers and orbital photographic mapping missions from 1966 through mid-1968. Their task, simply put, was to lay the scientific groundwork for the manned landing missions that were then being planned. The Soviets, flying both landers and orbiters under the Luna program name, successfully landed *Lunas 9* and *13* on the lunar surface and photographed their surroundings. *Lunas 10, 11, 12,* and *14* were placed in orbit about the Moon to map its surface and gravity field, providing gravity and landing-site data that would help insure against future debacles such as those that destroyed *Lunas 5, 7,* and *8.*

The American program for global mapping of the lunar surface and gravity field from orbit was called the Lunar Orbiter program. Thrown together on very short notice to prepare the way for manned missions, the Lunar Orbiter spacecraft borrowed heavily from military photographic reconnaissance satellite technology that Eisenhower had set in motion in 1958. The five Lunar Orbiter spacecraft, launched at three-month intervals in 1966 and 1967, were all outstandingly successful. Tracking of the Lunar Orbiter spacecraft provided a wealth of detailed data about the gravity field of the Moon's near side and the surface roughness in potential landing sites, data of critical importance to both American and Soviet manned missions.

The Surveyor series provided the American lunar program with experience in landing small automatic spacecraft on the Moon. Seven Surveyor landers were launched between May 1966 and January 1968. *Surveyors 1, 3, 5, 6,* and *7* worked beautifully, photographing, analyzing, and digging trenches in the lunar surface. By early 1968, both the Soviets and the Americans had collected the basic scientific and engineering data about the Moon that were needed to prepare for manned landing missions.

In 1968 both the Soviets and the Americans were also well advanced in testing the hardware that would be used for manned lunar missions. Sending manned missions to the Moon requires larger boosters than the "A-class" vehicles previously used in Earth-orbiting flights in the Soviet Vostok, Voskhod, and Soyuz programs or the Atlas and Titan ICBMs used in the American Mercury and Gemini programs. The American space program, led by Wernher von Braun's design team at Redstone Arsenal built the *Saturn 1* for testing Apollo lunar-mission hardware in low orbit about Earth and the huge *Saturn 5* for launching manned lunar-landing missions. The Soviets adapted the Proton ("D-class") heavy-lift vehicle to test lunar hardware in Earth orbit. For manned lunar-landing missions, an even larger booster was required. This behemoth "G-class" booster was the key to Soviet victory in the race for a manned lunar landing.

Three modified man-rated Soyuz spacecraft, called Zond ("probe" in Russian), carried aloft on the Proton heavy-lift vehicle, were fired on trajectories passing around the Moon and back to Earth in March, September, and November of 1968. No lunar-landing hardware was, or could be, carried on these flights, owing to the limited payload capacity of the booster. Simultaneously, American training flights using the *Saturn 1* were carried out in low Earth orbit, testing the hardware and procedures to be used in the lunar-landing mission.

Meanwhile, American spy satellites monitored the progress of huge new launch facilities under construction at the Tyuratam launch facility in Kazakhstan, where all Soviet manned flights originate. The two G1 launch pads were apparently structurally complete by August 1966, and all signs of construction and testing were cleared away by May 1967. Sources in the intelligence community say that a G1 booster was on the pad for several months in early 1968, but no launch was attempted, possibly due to difficulties experienced in static testing of the giant first stage. Both pads then sat empty for about a year. A new booster was installed on one pad in April 1969 and was prepared for launch. On July 4, 1969, the G1 booster exploded on pad number two, demolishing itself and the pad. (Later launch attempts in June 1971 and November 1972 ended with inflight explosions. But by then, the race to the Moon was already long over.)

In December 1968, the three *Apollo 8* astronauts rode a *Saturn 5* out to the Moon and into orbit about it. The Soviets at that point immediately canceled plans for further Zond flights around the Moon. In retrospect, the story seemed clear. The Soviets knew in 1967 that their G-class superbooster was running behind schedule while the American *Saturn 5* was on schedule. The only possibility for the Soviets to land cosmonauts on the Moon before the Americans was if the *Saturn 5* encountered a serious (that is, fatal) problem in an early test flight. This possibility could not be discounted, but it seemed less likely as time progressed. However, there was one manned mission to the Moon that the Soviets could with reasonable confidence hope to do first: to send one or two cosmonauts in a modified Soyuz capsule on a quick loop around the Moon and return them to Earth. That is precisely the capability that was demonstrated by the unmanned Zond missions of 1968. This flyby mission could be flown on the reliable Proton booster and could be done quickly. By this plan, had the September 1968 flight gone flawlessly, the November flight would have been manned. The first person to "visit" the Moon (albeit only on a flyby mission, not a landing) would be a Soviet cosmonaut. This would certainly take the edge off the impending Apollo lunar landing. Further, the first manned Apollo mission to the vicinity of the Moon was scheduled for early 1969, giving time for both November and January flight opportunities for Soviet missions before the first American mission. But the September flight was not flawless: the spacecraft reentered Earth's atmosphere at so steep an angle that the deceleration would have killed a human passenger. Under the rigorous safety rules of the Soviet manned space program, this meant that the November mission had to be flown without a crew to demonstrate the flight profile before a cosmonaut could be flown. If all went well, the first manned flight would then be attempted in January 1969. But at almost the same time, NASA announced that the Apollo program was running so far ahead of schedule that the first manned *Saturn 5* mission, *Apollo 8*, would be launched in December on the first complete *Saturn 5* launch vehicle. The mission: fly three astronauts without the basic lunar landing hardware out to the Moon; orbit the Moon for several days, just as the command module would be required to do on a real landing mission; and then return to Earth. When that mission returned successfully to Earth, the Soviet "end-run" program to launch cosmonauts on Zond was canceled as being too little, too late.

On the heels of the *Apollo 8* success, NASA scheduled a full dress rehearsal for a lunar landing, with all the landing hardware exercised, on the *Apollo 10* Moon-orbiting mission in May. The first manned landing was set for *Apollo 11* in July. The Soviets tried to upstage this event in two ways: first, the *Luna 15* "mystery spacecraft," quite possibly an attempt at automated return of a small sample of lunar material, was dispatched to arrive at the Moon while *Apollo 11* was in flight. Second, the G-class superbooster was set for launch on July 4, calculated to take the edge off American Independence Day festivities. But *Luna 15* crashed and the G1 exploded—and *Apollo 11* made history. The race was over. The Soviet lunar program retreated to unmanned missions, featuring automated retrieval of tiny samples of lunar soil by the *Luna 16, 20,* and *24* missions.

By the end of 1972, the *Apollo 11, 12, 14, 15, 16,* and *17* missions had all landed on the Moon and returned safely to Earth. Unlucky *Apollo 13*, after suffering a nearly fatal explosion of an oxygen tank outbound from Earth, defied all odds by returning its crew safely to Earth. Each landing brought back a load of precious lunar samples. Twenty-seven men orbited the Moon, and twelve astronauts, all American, walked the lunar surface. But then Earth lost interest in the Moon. From 1976 to 1994, the only mission to visit the Moon was a small spacecraft launched by Japan. Further, no nation on Earth presently has the ability to launch manned lunar missions. The eunuchs ordered the fleet destroyed.

In 1994 the eighteen-year hiatus in American and Soviet lunar exploration was broken by a most unlikely newcomer. The Clementine spacecraft, using technology developed by the Reagan-era Strategic Defense Initiative Organization (SDIO), contains a variety of cameras and sensors in a small (three-hundred-pound) high-tech package, complete with a powerful rocket engine. Clementine is a civilian outgrowth of SDIO's Brilliant Pebbles program for interception of ballistic missiles, a small, sophisticated plowshare fashioned from the last sword of the cold war. Launched by a reconditioned Titan 2 ICBM retired from active service with the Air Force, Clementine was dispatched to the Moon in January 1994. There it orbited for several months before minor problems precluded the planned extension of the mission outward from the Moon to a flyby of a near-Earth asteroid. During its months of active study of the Moon, Clementine produced stunning compositional and photographic maps of the Moon, even collecting fascinating evidence suggesting the presence of massive deposits of ice at the lunar poles. Perhaps the most impressive achievement of Clementine, however, was the price tag—several times less than the cost of a NASA mission with comparable capabilities. At least one follow-up Clementine mission to the Moon and nearby asteroids is already in the planning stages.

These lunar missions provided a wealth of information about the Moon's nature and history. The largest portion of our present knowledge comes from the study of returned lunar materials in laboratories on Earth. Of course, the overwhelming preponderance of the available lunar material was the half ton returned by the Apollo astronauts, who also carefully documented the locations of the returned materials. The relatively tiny Luna samples, however, have an importance out of proportion to their mass, because they sample three separate and independent areas of the lunar surface not otherwise studied. For safety reasons, most of the landings for these sample retrieval missions were planned for flat, smooth terrain. This clearly biases the samples in favor of mare materials. It was no surprise to find that the rock chips found in the lunar regolith (impact-pulverized surface material) and the large single rocks picked up on the surface by astronauts are mostly basalt and closely similar materials. Both Apollo and Luna samples have been made available to numerous qualified investigators in many nations. In a very real sense, the study of the Moon has been an international endeavor.

These samples have in recent years been supplemented by several small meteorites that were found on the ice in Antarctica along with over ten thousand other more conventional (asteroidal) meteorites. These few rocks, which differ significantly from other meteorites in age and composition, are clearly pieces blasted off the Moon by the impact of comets or asteroids on the lunar surface. Although they surpass the Luna samples in mass, they suffer from the fact that we have no idea what area of the lunar surface they come from.

The entire near side of the Moon has been mapped by Earth-based spectroscopy. Systematic differences between the highlands and mare basins are evident, but it is not easy to analyze these spectra in terms of the spectra of individual pure minerals, or even the spectra of meteorite classes. The reason for the difficulty became clear with the return of the first Apollo samples: the lunar surface is not composed of a simple mixture of minerals. Recall that minerals are, by definition, naturally occurring crystalline solids. Ice is a mineral; crude oil is not. Glass, which is very common in the lunar regolith, is actually a disordered, supercooled liquid that was cooled too quickly to crystallize into mineral grains. The sharp, distinctive absorption features seen in the spectrum of each pure mineral owe their distinctness to the very regular, highly ordered crystal lattice of that mineral. Glass, on the other hand, has no regular lattice structure: it is therefore not a mineral. Accordingly, the features in its spectrum are severely broadened and blurred. For this reason, interpretation of the spectrum of glass is much more difficult than those of the pure minerals out of which the glass was made.

The glass in the regolith owes its origin to impacts of cometary and asteroidal material with the lunar surface. The enormous kinetic energy (energy of motion) of the impactor given off in the instant of impact crushes, excavates, melts, and vaporizes lunar material amounting to roughly 100 times the mass of the impacting body. Some of the vapor condenses as tiny liquid droplets that cool rapidly. Larger droplets of impact-melted rock, hurled violently from the crater, splash onto the lunar surface nearby while still in the liquid state, where they stick to and weld together mineral grains as they quickly cool to form glass. These welded lumps of minerals and glass are called agglutinates. The longer a region of the lunar surface is exposed to the celestial rain of impacting material, the larger the abundance of glass and of glass-welded agglutinates. Regolith with a high glass abundance is said to be "mature." Agglutinates greatly complicate the business of extracting useful materials from mature lunar regolith. They weld together mineral grains of radically different compositions and properties, rendering most techniques of mineral separation and enrichment ineffectual.

The Apollo and Luna samples give us an opportunity to determine "ground truth": we can see precisely how the reflection spectrum is related to the composition and the mineral and glass content of the surface. We may then have much more confidence in our ability to interpret the spectra of not-yet-visited regions of the lunar surface.

In general, lunar rocks differ from terrestrial rocks in that the Moon contains much lower concentrations of easily vaporized elements. Volatile elements such as hydrogen, carbon, and nitrogen are much scarcer than in Earth's crust. Further, even elements that do not become volatile until heated to a thousand degrees Celsius, such as sodium and potassium, are distinctly depleted on the Moon. There is a corresponding enrichment of very involatile (refractory) elements in the lunar surface. These refractory elements include calcium, aluminum, and titanium.

Given Earth-based spectra and returned samples, we now know that the mare basins are filled with deep basaltic lava flows that have been reprocessed by the rain of impactors into regolith many meters deep. The individual mare basins differ significantly in composition. Mare basalts differ from the typical terrestrial basalt in two important respects. First, the lunar basalts are much less oxidized than terrestrial basalts. Iron can exist in the unoxidized state as metallic iron, as in iron meteorites or Earth's core, in a moderately oxidized state as ferrous iron, with a charge of +2 on each iron ion, and in a highly oxidized state as ferric iron, with a charge of +3 on each ion. Ferric iron compounds, which are generally red, are commonly and readily produced on Earth, where the presence of water and oxygen encourages oxidation of iron to make the mixture of ferric minerals we refer to as "rust." The most common component of rust is the red ferric oxide that constitutes the mineral hematite, whose name is derived from the Greek word for "blood." On the waterless, airless, oxygenless Moon, ferric compounds are almost impossible to make. Thus, ferric iron is common in terrestrial basalts, which have been repeatedly recycled through Earth's highly oxidizing surface environment, but absent in lunar basalts. On the Moon we find ferrous, not ferric, minerals and glasses, sometimes even accompanied by tiny traces of metallic iron.

The second distinctive feature of lunar basalts is that many of them have a very high titanium content. Some mare basalts have 15 percent to 20 percent titanium, mostly in the form of the mineral ilmenite, compared to 1 percent or 2 percent on Earth. Lunar ilmenite is an iron titanate, containing only ferrous iron, titanium, and oxygen, in the proportions $FeTiO_3$. Studies of the spectra of lunar mare basins by Bob Singer of the University of Arizona have produced detailed maps of the distribution of iron and titanium in the mare regolith. These maps, calibrated to agree with the composition of the returned Apollo and Luna samples, serve as an excellent guide to the minerals and chemistry of the nearside mare basins.

But the mare basins make up only a quarter of the Moon's near side and only about 2 percent of the far side. Nearly seven-eighths of the lunar surface is covered by highlands. The highlands are made not of basalt but of rocks with much higher calcium and aluminum contents, and much lower iron and magnesium contents, than any basalts. The principal highland rock is anorthosite, in which the dominant mineral is anorthite, a calcium aluminum silicate with the proportions

$CaAl_2Si_2O_8$. The highlands consist principally of a thick, nearly global crustal layer of low-density, refractory-rich rock. The mare basins, by comparison, are holes blasted through the anorthosite crust by comet and asteroid impacts, then flooded by dark, dense basalt lava from below. Because the basalt liquid, with its high content of the relatively heavy elements iron and titanium, is so much more dense than anorthosite, the basalt lava flows can flood only the lowest parts of the anorthosite crust. The Oceanus Procellarum seems to be just this kind of feature.

Although anorthosite and basalt are the dominant rocks on the Moon, significant amounts of several other related rock types occur. One of these, called KREEP, is sufficiently distinctive to merit a few words of explanation. The name itself is a reminder of its distinctive chemistry: the letters stand, respectively, for potassium (K), the rare earth elements (REE), and phosphorus (P). The rare earths are a family of very rare heavy elements with such similar chemical behavior that they tend to travel together during geological processing. KREEP is unusual for its very high content of phosphates of the rare earths. Among the elements that follow the rare earths are the radioactive elements uranium and thorium. But potassium itself is one of the most important radioactive heat sources in the planets. A minor isotope of potassium, the one with atomic weight 40 (^{40}K, read as potassium-40), is radioactive. Thus, KREEP happens to be a highly concentrated extract of the three most important radioactive materials in the Moon—uranium, thorium, and ^{40}K. Here is a clear example of a rare rock type that has been strongly concentrated by lunar geological activity. If there were a viable economic use for KREEP, we would call it an ore. Its presence warns us that unfamiliar types of ore-formation processes may operate on the Moon in the absence of oxygen and water. We must keep an open mind about the possible presence of ores.

The regolith at any point on the Moon is, not surprisingly, dominated by crushed fragments of local rocks. But impact events can eject rock chips to great distances from their point of origin. Sufficiently large impacts can hurl crater ejecta to any point on the lunar surface. Thus, a large sample of regolith from any point on the Moon contains numerous tiny rock chips from major craters made anywhere on the Moon. For example, regolith samples from Apollo's mare-basin landing sites contain anorthosite rock chips from the highlands. Any large sample of the surface from a single point samples a great number of other points as well, but of course we do not know the source of any particular piece.

But there are certain peculiarities of the composition of the regolith that point to other influences. For example, the regolith contains about 100 parts per million (100 grams per ton) of metallic iron-nickel alloys. No such material can be found in native lunar rocks, but it is a common constituent of meteorites, which are for the most part simply fragments of asteroids. Clearly, the iron-nickel metal on the surface of the Moon is contributed by asteroid impacts. During impacts, tiny fragments and droplets of asteroidal materials, including metal, are sprayed across the face of the Moon along with the lunar material ejected from the crater.

In addition, the mature regolith contains small amounts of many volatile elements, especially hydrogen, carbon, and nitrogen. Although the fresh local rock has only the tiniest trace of these elements, commonly less than one part per million, the regolith has higher concentrations *in proportion to its maturity.* Since maturity means higher glass content, it seems reasonable to suspect that the glass may be the host of the volatiles, but such is not the case. Instead, almost all of the volatiles are found embedded in the surfaces of ilmenite grains. The smallest ilmenite grains, being "all surface," have the highest proportions of volatiles. The largest ilmenite grains, if treated with a chemical agent that dissolves only a thin layer from their surfaces, give up virtually all of their hydrogen, carbon, nitrogen, and other gases. This is practical, useful information, since it makes extraction of these gases relatively easy. But how can we account for their peculiar distribution?

The source of these embedded gases is the solar wind, the stream of ions of hydrogen and other gases expelled into space at speeds of hundreds of kilometers per second by the Sun. These solar gases contain a large amount of helium, the second most abundant element in the Sun. Solar helium is a mixture of two isotopes—normal (heavy) helium with an atomic weight of four units (^{4}He), and a trace of light helium (^{3}He).

Helium is also made by radioactive decay of uranium and thorium, both of which decay by emission of alpha particles. These alpha particles, fired out by the decay of a heavy nucleus within a mineral grain, collide with many atoms in the mineral lattice and coast to a stop not far from their point of origin. Alpha particles are just high-energy ^4He nuclei; after decelerating and picking up electrons, they become atoms of ^4He. This helium produced by radioactive decay, called radiogenic helium, consists of pure ^4He. It is located throughout the grains of minerals that contain traces of uranium and thorium, not on grain surfaces.

The solar wind is deflected from impact with Earth by our planet's magnetic field. The small residue of solar wind that successfully penetrates the magnetic field by trickling in at the poles serves to excite auroral displays at high latitudes, but it is stopped from penetrating to the surface by Earth's dense atmosphere. On the Moon, there is no strong planetary dipole field and no atmosphere of any consequence to block the solar wind particles. Instead, they slam into the surface at full speed, penetrating about a ten-thousandth of a centimeter into any solid material on the surface before coming to a stop. The decelerated, implanted gases then can leak rather efficiently out of most minerals except ilmenite, which, because of arcane details of its crystal structure, is better than other common lunar minerals at retaining small atoms of foreign gases. The concentration of solar-wind gases in the local regolith therefore increases with the age of the surface, achieving the highest levels in regions with very high ilmenite contents.

An engineer intent on extracting hydrogen and helium from the lunar regolith with the highest possible efficiency would use some physical separation technique to enrich ilmenite, then screen the ilmenite mineral grains to eliminate all except the very smallest grains. These ilmenite-rich "fines" accumulate very high concentrations of all solar-wind gases when they are present in mature soils. We return to the economic significance of these gases in chapters 4 and 8.

In addition to the asteroidal iron-nickel grains in the regolith, there are also tiny amounts of nearly pure iron metal in some lunar rocks. The provenance of these rare grains is interesting. When a sample of regolith is melted by an impact event, it will generally contain both ferrous-iron-bearing silicate minerals and a small amount of hydrogen from the solar wind. The hydrogen reacts with the ferrous iron in the melted rock, extracting some oxygen from the melt to make water vapor and metallic iron. The water vapor is quickly lost into space because of the Moon's weak gravity, and the atoms of metallic iron accumulate to make tiny droplets of liquid iron. If the rock is held liquid for a long time, the droplets of dense liquid iron will agglomerate and eventually settle out of the liquid rock. But impact melt usually cools too quickly for such separation to occur. The result is a fine-grained rock containing small traces of metallic iron. The nickel content of the original rock is so low that the metal formed in it is virtually pure iron, very unlike the iron-nickel-cobalt alloy found in most asteroids.

Asteroids are not the only bodies to impact on the Moon: comets of all sizes also strike the surface. Toward the larger sizes (several kilometers in diameter), comets are actually more abundant than asteroids. Comets differ from asteroids in composition in that comets contain abundant water ice, and possibly other ices as well. Comets, which pursue very elongated orbits about the Sun, may be traveling at speeds as high as seventy-three kilometers per second as they approach the Moon. Their impact energy per gram increases with the square of their relative velocity. Compared to an asteroid of equal mass traveling twelve kilometers per second relative to the Moon, a cometary impactor will carry thirty-six times as much kinetic energy. Comets therefore tend to explode into vapor with near perfect efficiency and devastating power on impact. The vapor cloud from a comet impact explosion will be so hot and fast-moving that it will not only escape almost completely from the Moon but also blast an even larger mass of lunar material away into space. The net effect of comet impacts is to erode the Moon, not add mass. Nonetheless, some tiny proportion of the water and other volatiles in comets will be left behind on the Moon as evidence of the impact. That water may condense in the very coldest regions of the Moon, in permanently shadowed crater bottoms near the lunar poles, to form deposits of ice.

All schemes for the utilization of lunar material depend on our knowing not only what is present on the Moon but also how to concentrate, separate, and process those materials into useful products. This lore is in part based on familiar terrestrial experience, but for the most part it is a novel adaptation of familiar principles of chemistry and physics to a truly alien environment. Normal terrestrial mining, processing, and fabrication technology often proves to be disastrously inappropriate on the Moon, Mars, or an asteroid.

Just as the first animals to emerge from the ocean a half billion years ago had to learn new abilities and techniques to live on land, so we must learn many new tricks to survive on the islands of the cosmic sea to which we embark. Our remote ancestors took two hundred million years to learn how to adapt to the land. We, however, propose to make the transition to life in space in a single generation. But we are no longer limited by the glacial rate of natural genetic innovation. Instead, we rely on technology. Indeed, even human biological adaptation to some alien environments may be brought about not by random mutations and natural selection but by conscious genetic engineering. This is the age of technology. To predict the future without reference to rapidly evolving new technologies is to underestimate the possible—and to lend credence to the counsels of despair.

4

CASTLES BUILT
OF MOONDUST

A young technological civilization benefits from arising on a world that has a nearby satellite to serve as a challenge and a first target for spacefaring. But it benefits even more if that satellite is harsh and unforgiving, motivating an extension of exploratory activities to other, more distant, but more rewarding targets. If the ancient records can be believed, humanity originated on just such a world.

—Encyclopedia Galactica
From the entry, "Age of Exploration"
13382 CE

We cannot instantly determine whether a return to the Moon can be justified on economic grounds. The road to that decision point is a long one, traversing great expanses of science, technology, and economics. Along that road, there are several natural milestones in the study of any solar system body. The first step is remote observation by observatories on Earth or in orbit above Earth's atmosphere. The second step, which became possible to us in approximately 1960, is to send an instrumented flyby probe to pass by that body at high speed and telemeter its observations back to Earth as radio and television signals. We have so far reached this stage with every planet except Pluto, including the natural satellites of these planets, and also several comets and asteroids. The third step is usually to dispatch a hard-landing impact probe to study the atmosphere (if any) of the body, to measure its gravitational and magnetic fields, and to examine the surface (if visible), in preparation for more advanced exploratory missions. Entry or hard-landing (kamikaze) probes have been sent to the Moon (Ranger) and Venus (Venera and Pioneer Venus). Several Soviet Mars missions have inadvertently found themselves in this category.

The fourth step is to place an instrumented spacecraft in orbit about the body by judicious firing of a rocket motor on a flyby vehicle as it passes by. Such artificial satellites have been placed in orbit about Earth, the Moon, Venus, Mars, and Jupiter to develop global maps and study time-dependent phenomena such as planetary meteorology. Notable examples include the Pioneer Venus Orbiter, the *Mariner 9* and Viking Orbiters at Mars, and the *Galileo* orbiter at Jupiter. The Cassini mission currently under preparation is intended to orbit Saturn.

The fifth step is to drop survivable probes into the atmosphere or onto the surface of a body. A number of survivable landers have been sent to the Moon, Venus, and Mars, including the *Luna 9* mission and the two Viking Lander spacecraft on Mars. The Galileo mission has dropped an atmospheric entry probe into Jupiter, and the Cassini mission will drop the Huygens entry probe onto the surface of Saturn's huge moon Titan.

Once survivable landers are possible, landers with mobility are the logical sixth step. These may take the form of balloon-borne atmospheric floaters, as in the Soviet-French Vega missions to Venus, or automated surface rovers, as in the *Lunokhod 1* and *2* rovers landed on the Moon by the Soviet *Luna 17* and *21* missions. Mars rovers are planned for future international collaborative missions. Automated sample-return missions are a natural extension of this category.

By this stage, the groundwork necessary for manned missions has been accomplished. Manned landing missions may now be sent to bodies on which human landings are possible and feasible. The only body in the solar system for which such seventh-step missions have been attempted is the Moon. Those missions were the Apollo landing missions described in chapter 3. Manned expeditions to Mars have been widely discussed, but present mission concepts are so expensive that no nation has made plans to build and launch such a mission. Manned visits to the Martian moons Phobos and Deimos often figure as parts of manned missions to the surface of Mars.

But this is by no means the last possible step. Beyond brief exploratory landings, the next (eighth) step would be the establishment of a long-term surface research station or base. The construction of a base on the Moon has been a recurrent theme in both the American and Soviet manned-spaceflight programs, and it has been extensively discussed in Japan for the past decade. The need to protect the residents of the base from environmental hazards places severe demands on its design and construction. Even in the early stages of such a base, astronauts would visit for months at a time. It is obviously essential to provide the crew with air, water, and food, but it is much less obvious what the best means for achieving these ends might be. For short exploratory missions of a few days, it is clearly most expedient to carry along the requisite supplies from Earth. For lengthy missions, recycling and regeneration of air, water, and food become increasingly attractive. At what point does a closed ecological life-support system, with the ability to grow a variety of basic foods, become the "best" solution? And how could such a system be built, stocked, and maintained? Given what we know of the Moon's resources, is such a system workable?

Long stays on the lunar surface demand prolonged exposure to an environment differing substantially from that at the surface of Earth. For example, the gravity field on the lunar surface is about one seventh as strong as on Earth. The effects of years of exposure to such low gravity fields, while not likely to be catastrophic to human health, are nonetheless very poorly understood. Prolonged experiments in reduced gravity cannot be carried out on Earth. Even in a space station, such intermediate gravitational accelerations can practically be achieved only in centrifuges. We might find that, after lengthy stays on the Moon, astronauts suffer debilitating degenerative disorders. More likely, we might find that they adapt readily to lunar conditions of weak gravity and thrive in that setting, but become increasingly unable to cope with normal terrestrial gravity. Or there might be no important long-term negative effects due to low gravity. The 1995 flight of the American physician Norman Thagard to the Russian Mir space station points the way toward an especially promising attack on the problems of weightlessness: bring superior Russian space-station capabilities and experience and superior American biomedical research instrumentation and techniques to bear on a common problem.

Another very important factor is long-term exposure to solar and cosmic particulate radiation. Cosmic rays, which are very high energy nuclei of light atoms, mostly protons, have energies ranging from millions to billions of volts. Life evolved on Earth with the radiation protection afforded by about one kilogram of air per square centimeter of Earth's surface. Substantially smaller amounts of shielding will undoubtedly admit much larger fluxes of harmful radiation. Long-term residents of a lunar base or space station would require shielding on the order of one kilogram per square centimeter to reduce their radiation exposure to the levels they would experience at sea level on Earth's surface. Of course, some populations on Earth have lived for many generations at high altitudes with substantially less shielding (but they did not know it was a problem and did not have to get advance permission from OSHA). There are also several simple and effective dietary countermeasures to moderate radiation damage, such as maintaining high levels of natural antioxidants and free-radical scavengers such as vitamins E and C. And eventually synthetic drugs can be designed to help combat the highly destructive free radicals made by cosmic rays. Perhaps reduction of shielding to 500 grams per square centimeter can be tolerated with no thought to countermeasures, and as little as 100 or 200 grams per square centimeter (12 to 24 centimeters; 5 to 10 inches of iron shielding) may be acceptable with appropriate countermeasures. Many fancy materials have been suggested for shielding astronauts against cosmic rays, but plain old water and ice are very near the top of the list in effectiveness per gram. Research on the radiation problem can be done anywhere in space, such as at the Russian Mir space station in orbit about Earth just above the atmosphere. But in fact little such work has been done, and the issue remains in doubt. America's persistent failure to build and use a space station to solve these problems will haunt us in the future. The vast Soviet experience with long-duration flight in space stations has not solved the problem either, because of the Soviets' weaknesses in theoretical space medicine and medical instrumentation. A continuing joint Soviet-American research program on these problems of gravitational biology and cosmic-ray protection would permit a fruitful wedding of the nations' individual strengths. Serious work on the radiation problem could begin almost at once, now that the arbitrary political barriers in the way of such collaboration have been removed and the Space Shuttle's compatibility with Mir has been demonstrated.

Design and construction of a lunar base can proceed even without a mature understanding of cosmic-ray effects: all we need to do is design conservatively and provide at least as much shielding as the atmosphere provides on Earth. It is simply infeasible to build and launch from Earth heavily shielded lunar-base modules, ones with a thousand grams of shielding per square centimeter. Consider a simple "telephone booth" storm shelter with a floor area of one square meter and a height of two meters. The shielding needed to protect this modest volume would weigh about 100 tons, equal to the full payload capacity of a *Saturn 5* to low Earth orbit (a few hundred kilometers up). It would take five *Saturn 5* launches to deliver a single such storm shelter to the Moon. Even if we still had *Saturn 5*s, this would entail an absurd expense. With only the much less capable Space Shuttle, it would take about twenty shuttle launches (plus new, specially designed upper stages—the Space Shuttle is incapable of flying to the Moon, let alone landing and returning!) to deliver the same weight. The price tag would be an absurd $10 billion. Clearly, shielding an entire lunar base adequately with shielding lifted from Earth is impossibly expensive. We are forced to plan a lunar base in such a way that it can be shielded by lunar materials.

Most discussions of space vehicles and stations contain lengthy discourses on the hazards of micrometeoroid impacts. But once we have provided our base with a carapace thick enough to provide protection against cosmic rays, we have automatically provided a highly effective meteoroid shield. We should keep in mind that delicate items of equipment exposed directly to the space environment will inevitably experience small impacts from time to time. Occasional maintenance and repair work will be necessary to keep large solar-cell arrays in top working order. But the probability that an astronaut in a radiation-shielded base would be killed or injured by a meteoroid impact is comparable to the probability that a person on Earth's surface will be struck and killed by a meteorite. Such events, described in my book *Rain of Iron and Ice,* are rare enough to be of little concern to individuals.

Any lunar base must be supported by a transportation system that fills several essential functions. First, the crew, lunar-base modules, and equipment must be launched into a low orbit about Earth. This first step—fighting against Earth's powerful gravity and against air friction during ascent—is the hardest, most expensive, and most hazardous. The Moon-bound cargo must then be carried from this low "space-station orbit" out to the vicinity of the Moon, where the cargo must be soft-landed precisely at a preplanned landing site on the Moon. Next, the base modules and equipment must be deployed and assembled on the lunar surface and covered with radiation shielding. Finally, crew members being rotated back to Earth, along with samples of lunar materials and other cargo, must be lifted off the lunar surface and sent on a trajectory that returns either to the space-station orbit or to the surface of Earth. Of course, every piece of lunar-transportation hardware that returns to Earth's surface must be replaced by launching a new vehicle from Earth at great expense. But vehicles that simply shuttle back and forth between a space station and the Moon need not be replaced, only refueled and maintained.

Such a transportation system will exert a constant demand for fuel both at the space station and on the lunar surface. If this fuel all originates on Earth, the cost of launching and transporting it out to these remote "gas stations" becomes a large part of the cost of the overall mission. But if fuel could be made on the Moon, the transportation system would be much more economical. At present prices, it costs about $4,000 per pound to launch propellants (or drinking water!) into low earth orbit—except with the Space Shuttle, in which case the cost is more like $10,000 per pound, or $80,000 per gallon! Fortunately, there are a variety of known ways to lower cargo launch costs by a factor of five, ten, or even twenty. Even so, the need to recycle water and other materials is obvious. But such economic considerations have had no significant effect on the way government-run launch vehicles are designed and operated. Governments, unlike industries, do not have to compete and show a profit!

It is certainly not possible to foresee a need for a permanent lunar colony at this time, and we shall not start with any philosophical or political arguments to the effect that such lunar colonies are either desirable or inevitable, let alone profitable. Instead, we shall examine the resources of the Moon and the known technologies for using lunar materials to see whether such a colony might make logistical and economic sense as a self-sufficient entity. If so, then further consideration is sensible. If not, then such a colony would be nothing but an expensive stunt or hobby, a means of consuming, not expanding, humans' available resources. However, if the development of the Moon should progress to the ninth stage with the establishment of a self-perpetuating colony, then all of these needs for life-support materials, radiation shielding, and propellants would be amplified. It would become even more desirable to manufacture as much as possible of the colony's needed items out of lunar materials, in order to minimize the need for expensive launches from Earth.

The task of designing in a single step a self-sustaining economy for a lunar colony, starting (as we now are) with nothing, is too ambitious by far. We need to begin with the earlier steps in lunar exploration and the simplest schemes for lunar resource use to see whether there is any plausible way to get our foot in the door. If not, the rest is pure fantasy.

What, then, do we need on the Moon? In what order do we need these resources, and how much do we need? How might we extract, process, and fabricate useful products from them most efficiently? At how early a stage might lunar resource use be integrated into a program of lunar exploration?

Humans have already carried out the first seven steps of lunar exploration, from telescopic investigation through manned lunar landings, with only a single, noneconomic, use of lunar materials: to return them to Earth as scientific samples. By characterizing this use as noneconomic, I mean simply that it was not done for monetary profit, and no profit was in fact realized. I certainly do not mean to denigrate the purely scientific interest in and value of these samples: as a planetary scientist, I am aware of the enormous importance of such samples in establishing the key events of solar system history and the essential processes of planetary evolution. Nor do I mean to imply that these samples have no long-term economic implications, or that we would have refrained from taking such samples if we were impelled by purely economic motives. As the codirector of a space engineering research center dedicated to the use of nonterrestrial materials, I am well aware of the powerful stimulus to thought and experimentation on lunar resource use provided by the availability of Apollo and Luna samples. I draw the distinction of economic motivation because *the time to do so has now arrived: we must address economic issues, or there will be no future space activity.*

In the following chapters, I emphasize what *can* be done, not what *will* be done. There is a sound scientific and engineering basis for defining capabilities. In some cases, an economic analysis is also possible. But governmental programmatic decisions depend on political factors that often have nothing to do with the logic of mathematics, chemistry, physics, engineering, and economics. The Apollo missions were done not because they were the next logical step in space (they were not), but because of a political decision to beat the Soviet Union to a major space landmark. Aerospace engineers often talk about "human factors" in spacecraft design, but the most profound human factor of all is the decision whether or not to put a program in the budget. Making that decision, alas, is an imperfect art, upon which the future of space exploration has long rested. But if economic factors are included, our choices of future missions will be dramatically changed. Rather than expending vast sums on political posturing, we may instead choose to invest in potentially profitable space enterprises. Any such scheme must be assessed "from the ground up." Actually, a better place to start might be *below* ground level. Our task is to provide a viable basement for the lunar infrastructure, upon which all lunar construction and economic activity must be built.

The Zen-like act of excavating a hole in the ground before beginning the construction of a tall building provides an apt metaphor for the beginning of construction on the Moon. As I have suggested, our proper task is to explore stepwise the contributions that use of lunar materials can make to future lunar activities, beginning with the installation of a small lunar base. That is not only the next logical step onto the

Moon but also the first step at which the use of lunar materials offers possible benefit.

The first step in installation of a lunar base will be to land the first base module in an appropriate location on the lunar surface. The module will, for the economic reasons previously outlined, be built and transported without radiation shielding. Therefore, the first task on the Moon will be to excavate a trench in the regolith deep enough to bury the module. The module will then be lowered into the prepared trench and covered by a few meters of loose regolith from the excavation, affording its future occupants adequate protection against solar flares, cosmic rays, and meteoroid impacts. A single doorway with an airlock, either in the trench at the end of the module or atop it, will provide access from outside the shielding layer. Virtually no processing and preparation of the raw regolith is necessary, except perhaps for the simple precaution of removing large rocks from the shielding mass to minimize the chances of inadvertent damage to the module during burial. The sole piece of equipment needed for the installation of this radiation shielding is a glorified shovel. It is literally possible for a couple of astronauts with shovels to do the job in a few weeks; however, astronaut time is notoriously expensive. A small piece of automated equipment would provide the capacity to excavate the trench and tuck in the first module and all subsequent ones. This is almost certainly the first practical use of lunar materials to be attempted.

Not all shielding materials are equal in performance. In terms of stopping power per gram, the ideal material for stopping cosmic-ray protons is hydrogen. Liquid hydrogen unfortunately not only has a very low density, necessitating very bulky hydrogen tanks, but it also is a "deep cryogen," boiling at temperatures only twenty degrees above absolute zero. On the Moon or in nearby space, constant refrigeration is required to maintain hydrogen in the liquid state without it boiling away. Fortunately, several hydrogen compounds, including water, liquid methane, and liquid ammonia, have very high concentrations of hydrogen atoms without the severe storage problems of liquid hydrogen. Of these, the most abundant and the easiest to handle and store is water. As noted earlier, water is the most useful shielding material against cosmic rays and solar protons. If we could, we would elect to use water or ice for radiation shielding of the lunar base. But water is in notoriously short supply on the parched lunar surface.

After burial, the first lunar base module is now ready for occupancy. A manned landing nearby will deliver the first few crew members, who enter through the airlock. Initially, life support will be provided as on a space station. Water, oxygen, and food will be brought along from Earth at a cost of several thousand dollars per pound. Wastes will be collected, bagged, and dumped. Carbon dioxide will be scrubbed from the air by a chemical, lithium peroxide, which releases oxygen as it absorbs carbon dioxide to make lithium carbonate.

As the base grows, the cost of transporting the necessary water, oxygen, and food becomes ever more prohibitive. There will be a strong incentive to at least recover and purify water for reuse. Next, there will arise a desire to recover oxygen, which has been metabolized into carbon dioxide by the crew. This will require a device to "crack" carbon dioxide into its components. The best scheme presently available is to use high-temperature gas-phase electrolysis of carbon dioxide. A hot ceramic membrane with an electrical potential difference across it splits carbon dioxide (CO_2) into carbon monoxide (CO) and an oxygen ion (O^-). The oxygen ion passes through the membrane and emerges on the other side as pure oxygen gas. This process recovers exactly half of the oxygen content. The carbon monoxide gas is then heated with a solid catalyst that helps it react with itself to make solid carbon and carbon dioxide. The carbon is removed, and the new carbon dioxide is cycled back through the carbon dioxide electrolysis unit. In that way, all the oxygen can be recovered, with solid carbon dust as the only byproduct. The carbon can then be discarded or used for some other purpose. The very same process, operating on water instead of carbon dioxide, makes oxygen and hydrogen gas.

High-temperature electrolysis uses a large amount of electrical and thermal energy. The most obvious source of the necessary electrical energy would be solar panels set out on the lunar surface. The thermal energy could be generated by electrical heating using the same energy source, or it could simply be collected by a concave mirror that concentrates sunlight onto the processing equipment.

These schemes permit efficient recovery of water and oxygen but leave the problem of a food source unanswered. An oversimplified but instructive point can be made by considering the fate of carbohydrates transported to the lunar base from Earth. Carbohydrates, including sugars and starches, are compounds of carbon, hydrogen, and oxygen having the general formula $(CH_2O)_x$. After passing through animal metabolism, the carbohydrates emerge oxidized by oxygen in the air into CO_2 and H_2O, a 1:1 mixture of carbon dioxide and water. If we now recover all the water and crack the carbon dioxide into carbon and oxygen, we have recovered the oxygen used in metabolism of the carbohydrates and achieved the transformation of each carbohydrate unit into a molecule of water plus an atom of solid carbon. Our life-support system now contains more water than it had previously, the same amount of oxygen as before, and a supply of waste carbon.

Of course, the human engine is not perfectly efficient. A significant fraction of the food we eat emerges as undigested or undigestible solid waste. This waste may be burned to make a mixture of carbon dioxide, water, and nitrogen. The water and oxygen can be recovered as before, and the nitrogen (from undigested proteins) can be added to the air as a fire retardant. Again, solid carbon is made as a byproduct. Eventually we would deplete our food supply and accumulate a lot of carbon dust, but if we are planning only a short mission, that is not a problem. (Alternatively, if we had more time and brought a greenhouse with us, we could use the carbon dioxide to grow plants and make oxygen, thus closing all the cycles of these materials.)

If leakage of gases from the base or inefficiencies in the base's processing of wastes leads to a decline in the oxygen supply, an apparatus similar (or even identical) to the carbon dioxide cracking unit can split water vapor into its component elements, hydrogen and oxygen. The waste hydrogen may be used as a rocket propellant, or it can be stored or even dumped into space if no better use can be found for it.

At some point, it may become feasible to grow food inside the base modules. Since most crops take several months to mature and few take less than a month, space agriculture becomes attractive only when applied to missions with durations of many months. Experiments with hydroponic agriculture in laboratories on Earth, such as Biosphere II, and in year-long flights aboard Soviet space stations have already shown that this can be done in practice. As a valuable fringe benefit, photosynthesis by these crops not only cycles carbon dioxide back into food but also releases oxygen into the atmosphere. It is not yet clear which are the best crops and agricultural techniques to use, or whether developing lunar agriculture will prove economically superior to transporting food from Earth. That humankind should have reached the mid-1990s without knowing how to grow food in space, how to deal with weightlessness, and how to counter the effects of cosmic radiation serves as a more powerful condemnation of the Soviet and American manned space programs from 1960 to 1995 than anything else a critic could say. Tsiolkovskii saw in 1903 that space agriculture would become a critical limiting technology. We should have paid better attention.

Soviet food-growing experiments in space aboard the Salyut space stations have demonstrated seed-to-seed cycles of small quantities of some food crops in zero gravity. American experiments on food production in closed life-support systems have been conducted only on the ground, some by NASA and some at the famous Biosphere II project, near Oracle, Arizona. The multiacre Biosphere II facility, dedicated to discovering how to make self-sufficient colonies in space, has been built and operated without any involvement of NASA, entirely from private funding. After several rounds of serious managerial and personnel disputes, the research task at Biosphere has recently been entrusted to Wallace Broecker, a respected scientist from Columbia University. Expertise gained in these studies will help us raise food and regenerate oxygen in future space habitats.

From the earliest days of the existence of the lunar base, return transportation to Earth is essential. After the first few months, regular crew rotation will commence. But even before the first scheduled return, provision must be made for emergencies. In the event of severe injury or illness, or in case of serious equipment problems in the base module itself (electrical outages, fire, environmental equipment malfunction, depredations by voracious regolith mice, and so on), a hasty return to Earth may become essential. A rocket capable of taking off from the lunar base and returning to Earth must be capable of a velocity increment of three kilometers per second. This requires the consumption of a mass of hydrogen and oxygen propellant approximately equal to the mass of the return vehicle, probably a minimum of nine or ten tons per flight. If the best chemical propellant combination, hydrogen-oxygen, is used, then the proportions would be about one ton of hydrogen per eight tons of liquid oxygen, the exact proportons needed to burn a hydrogen-oxygen mixture without waste.

As noted earlier, oxygen can in principle be recycled efficiently within a lunar base. However, there is no source of extra oxygen to expend as rocket propellant. We have also seen that hydrogen is made as a byproduct of the makeup of oxygen by electrolysis of water. We cannot simply take that hydrogen for use as a propellant because each ton of hydrogen used as propellant requires eight tons of oxygen to burn it. Thus, electrolysis of water produces hydrogen and oxygen in the same proportions as would be wanted for propellants, and it leaves no leftover oxygen to make up for any losses from the base. True, we could imagine transporting tanks of water to the Moon from Earth to be electrolyzed into rocket propellants using solar power. But this does nothing to alleviate the cost of transportation from Earth; indeed, it places the entire burden of rocket-propellant supply on launch from Earth, the most expensive option of all. Much more desirable would be to find a source of rocket propellants on the Moon, so that the expense of bringing them from Earth could be eliminated. This would substantially defray the cost of operating a lunar base.

Since eight-ninths of the mass of propellant is oxygen, a source of oxygen from the Moon would solve eight-ninths of the propellant transportation problem. But if we were to find water on the Moon, that would solve the entire problem: electrolysis of water provides eight-ninths oxygen and one-ninth hydrogen automatically.

Unfortunately, water remains unknown in materials returned from the lunar surface. Everything we have learned about the chemistry of the Moon suggests an extremely dry interior for at least the last four billion years. The most promising sources of water are external, not internal. Water vapor from the impacts of comets and water-bearing asteroids may indeed condense in very cold crater bottoms near the poles, but the Apollo and Luna programs provided no evidence capable of verifying the presence of polar ice. The polar regions of the Moon are very difficult to study from Earth, since it is by definition impossible to measure the reflection spectrum of a surface that is in perpetual darkness. Further, no instruments capable of detecting ice deposits were placed in polar orbit about the Moon during the Apollo era. Since 1972, lunar scientists have been planning for a small unmanned lunar satellite that would complete the mapping of the Moon's surface and its gravitational and magnetic fields and search for ice at the poles.

Unfortunately, for over twenty years it was impossible for NASA to get approval for this mission. The proposed NASA mission would have cost no more than 1 percent of what was spent on Apollo. It would have cost less than half as much as each one of the three completed but unused *Saturn 5* boosters that were abandoned at the abrupt termination of the Apollo lunar-landing program.

The Clementine mission of 1994, described in chapter 3, was flown for a fraction of the cost of the proposed NASA mission. It returned exciting evidence that suggests massive deposits of water ice in shadowed crater bottoms near the lunar poles. This inference has yet to be tested by landing missions near the lunar poles; however, it is now legitimate to consider the possible use of lunar polar ice in support of future activities.

Nonetheless, it is good to keep in mind that very few geochemists expect water to be found in lunar rocks. The only plausible sources are hydrogen implanted in the regolith by the solar wind, which is demonstrably present in small quantities, and polar ice, which remains a matter of vigorous debate and ongoing research.

The concentration of hydrogen in the regolith is typically about 2 to 60 parts per million, depending on the maturity of the soil (how long it has been exposed to solar wind and micrometeoroid bombardment) and the local abundance of the mineral ilmenite. The approximately 40 parts per million of hydrogen in average mature regolith can be extracted by heating the regolith to a high enough temperature so that the hydrogen gas can diffuse out of the ilmenite grains wherein most of it resides. But this requires such high temperatures that the hydrogen gas can partially react with the ilmenite. Basically, hot hydrogen extracts an oxygen atom from the ilmenite, making water vapor, metallic iron, and titanium dioxide (TiO_2). Each gram of hydrogen makes nine grams of water: a hydrogen concentration of 40 parts per million is equivalent to about 360 parts per million (0.036 percent) water.

Supposing the average thickness of mature regolith on the Moon is ten meters, this means that extracting all the hydrogen from the regolith and converting it completely into water would provide a layer of water one centimeter thick. To extract each gram of water requires heating 2,500 grams of regolith up to red heat.

Since the majority of the hydrogen resides in the 10 percent or so of the regolith that consists of ilmenite, it would make sense, first, to spend some energy on crushing the regolith so as to liberate individual grains of ilmenite from the rock chips and the glass-welded agglutinates that abound in mature soil. Then electromagnetic techniques could be used to extract concentrated ilmenite. This kind of concentration of ores using density, magnetic, or electrostatic properties is called beneficiation. Beneficiation can fairly efficiently separate out those larger grains of ilmenite that are mechanically "liberated" from other kinds of grains by crushing. Then only the ilmenite-rich separate need be heated. Suppose that 50 percent of the hydrogen is in ilmenite. Crushing may successfully liberate half of the mass of ilmenite as separate grains. The smallest ilmenite grains, which have the highest concentration of hydrogen, are the hardest to liberate from other minerals. Achieving efficient liberation of the smallest grains requires crushing them to a very small particle size. Unfortunately, very fine powders are extremely "sticky" because of "static cling," and they are difficult to separate into their component minerals. Electromagnetic separation may successfully isolate a powder with 10 percent of the total mass of the sample and 30 percent of the total hydrogen content. The amount of hydrogen recovered per unit of heating energy used will then be about three times as high as it would have been if the bulk, unprocessed regolith had been heated. But the recovery of hydrogen has dropped from 90 percent or 100 percent to only about 25 percent or 30 percent. This brings the water recovery down to about 0.01 percent of the regolith mass, an amount so small as to make it very improbable that water extraction can be done economically.

If ice is available at the site of the lunar base, then the situation is dramatically different. Recall that the only places on the Moon cold enough to retain ice for billions of years are permanently shadowed crater bottoms near the lunar poles. Ice would be less stable at the surface than at a depth of a meter or so because of the relatively frequent flash-heating of spots on the surface by micrometeoroid impacts. It is easy enough to imagine brushing away a meter or so of dry dust covering an ice deposit. The buried ice would presumably be in the form of a thick lens of regolith with its pore spaces filled with ice, a sort of extraterrestrial permafrost. Lenses of rather pure ice are conceivable, but more likely is a permafrost containing 10 percent to 30 percent ice. Water could be extracted by distilling or melting blocks of permafrost. One could imagine a high peak near the pole, and hence near the ice deposit, where the Sun shines permanently. (From that vantage point, the half-risen Sun is seen to circle the entire horizon once a month.) A simple steerable solar collector that always faces the Sun could provide all the energy needed for processing the ice. All this works well if the mine site and processing site are both located close to the lunar base—that is, if the base is located at one of the lunar poles. An interesting fringe benefit of a polar base site is the availability of water or ice for use as radiation shielding for the base. The total mass of available ice in the polar regions remains utterly unknown.

In other respects, the lunar poles are not an attractive site for the lunar base. The poles are in highland terrain, where the ilmenite concentration is normally very low. The polar regions are less accessible to spacecraft launched from Earth, and all landing missions to the poles incur a significant weight penalty due to increased propellant requirements. The poles also receive far lower fluxes of sunlight and have far lower mean temperatures than near the lunar equator, where the Sun is high in the sky. High latitudes also complicate the process of deriving electrical power from sunlight.

∴ ✳

Power is, in any event, a serious problem on the Moon. The greatest intensity of incident sunlight is of course found near the equator, where the Sun is directly overhead at local noon. The length of the day on the Moon is twenty-eight Earth days. Fourteen Earth days of sunlight alternate with fourteen Earth days of darkness. The power demands of a base tend to be a little higher during the lunar night because of increased demands for heating and lighting. It may be necessary to shut down processing equipment with high power demands at night when power is at a premium. During the lunar day, when sunlight is abundant, power can easily be generated by solar panels that convert sunlight into electrical power. But how would the base supply its nighttime power needs?

In principle, a large bank of chemical batteries could be charged up by excess power while the Sun is up, and the base could run off the batteries when the Sun is down. But the power demands of the base are so great that the mass of batteries needed to run the base for fourteen days is prohibitive. All these problems become more severe at higher latitudes, where the intensity of sunlight is lower, but vanish if there is a permanently illuminated peak at the pole.

The biggest single drawback of solar cells, then, is the need for storing two weeks' worth of power. One other suggested scheme for power storage, besides batteries, deserves mention: excess daytime electrical power could be used to spin up large, heavy flywheels. At night, the rotational energy of these flywheels could be tapped to run generators. High energy densities (kilowatt-hours of power per kilogram of storage equipment) require very strong designer materials that would, at least initially, have to be made on Earth and transported to the Moon at great cost.

Also, many problems arise from shutting down complex processing equipment for two weeks at a time. The thermal stresses associated with cooling of the equipment at shutdown and heating the equipment at restart are far more severe than those incurred during operation at constant temperature. The probability of equipment damage is therefore greatly increased. This scenario also requires that both peak processing-power and peak recharging-power demands occur at the same time, necessitating a power system very large in size and mass. But there is another, more subtle difficulty: the morale of a base crew that works at fever pitch for fourteen days straight, then finds itself with a fourteen-day "weekend." And of course the processing equipment is only is use 50 percent of the time, a poor investment of very expensive resources.

For these reasons, many engineers and mission planners have favored the use of nuclear power for the lunar base. They imagine a multimegawatt nuclear reactor operating continuously, powering processing equipment that also operates continuously. Although a reactor weighs more than the corresponding mass of solar cells, the reactor system is far more attractive because it has no need for massive power-storage systems, such as batteries. Of course, much of the mass of a small reactor is shielding—but the radiation shielding for the reactor can be a big heap of lunar regolith, not a mountain of lead carried from Earth at immense cost. Reactor safety during transportation is not a big problem, since the fuel rods need not be inserted into the reactor core until it is safely installed on the Moon. In a sense, the most severe drawback of space nuclear power is political: no matter how safe your reactor, someone will picket your launch site, file endless litigation against the launch, and even sabotage valuable equipment.

Another clever scheme for massive energy storage deserves consideration. I described (on pp. 67 and 70) how electrical and thermal power can be used to decompose water into hydrogen and oxygen. The same device, run backwards, reacts hydrogen and oxygen together to make water and generate electrical power. Such a device is called a fuel cell. In other words, if we had a good supply of water, we could use half of our available solar heat and electrical power during the daytime to break water apart into hydrogen and oxygen by electrolysis. At night the same apparatus, operating as a fuel cell, would "burn" hydrogen and oxygen to provide electrical power. The water produced would be condensed and stored for recycling. This same basic cycle could also operate on the interconversion of carbon dioxide into carbon monoxide plus oxygen.

All told, the last of these three options seems best. Only a small amount of new technology is needed, and that seems well based on existing laboratory experience. We could design and build a test prototype of such a system in a few months, and even flight-test a small version of it on the Moon within three or four years. Once we have a demonstrated, reliable source of power, then processing schemes that use that power on the Moon can be confidently planned.

The first industrial use of power on the Moon will probably be for the manufacture of propellants and life-support materials. We have seen how difficult it is to extract hydrogen and water from the Moon profitably. But oxygen is another matter entirely. Hydrogen makes up perhaps forty parts per million of the lunar regolith, but oxygen is the most abundant element in lunar rocks. In the iron-magnesium pyroxene silicates of the mare basalts, 60 percent of the atoms are oxygen. In the anorthite of the lunar highlands, oxygen makes up 61.5 percent of the atoms. In ilmenite, oxygen accounts for 60 percent of the atoms. Because oxygen atoms are lighter than the metal and silicon atoms that make up most of the rest of the surface, oxygen makes up a somewhat smaller proportion of the surface on a weight basis, close to 40 percent. That is about ten thousand times the concentration of hydrogen. But this oxygen is bound in oxides of silicon, magnesium, iron, calcium, aluminum, titanium, and many other less common elements. Many of these oxides are so stable that it is extremely difficult to get the oxygen out of them economically.

Of all the abundant metals found as oxides on the lunar surface, the one that surrenders its oxygen most readily is iron. Iron oxide occurs most commonly in two of the most important minerals in mare basalts—ilmenite, with the formula $FeTiO_3$, and pyroxene, $(Fe,Mg)SiO_3$. The easiest way to get oxygen out of these minerals is to react them with an easily oxidized gas (what chemists call a reducing agent), such as hydrogen or carbon monoxide. Just as in the natural reduction of ilmenite by hydrogen in heated regolith, the products include metallic iron and oxides of the other elements. When hydrogen is the reducing agent, water vapor is the prime product. When carbon monoxide is the reducing agent, carbon dioxide is produced. It is also possible to use methane (CH_4) as the reducing agent, in which case both water vapor and carbon dioxide are made as products.

Extracting the oxygen from water vapor or carbon dioxide is the task of a high-temperature electrolysis unit such as the one just proposed for oxygen recovery in the life-support system of the lunar base. The hydrogen (or carbon monoxide) left behind is then recycled through the mineral-reduction process to make a fresh crop of water vapor, and so on repeatedly around the cycle. A little hydrogen will be lost due to the diffusion of hot hydrogen gas through hot metal, but otherwise the cycle is nicely closed.

Of the many proposed schemes for extracting oxygen from lunar rocks, several have been tested in the laboratory on simulated lunar materials. One of these proposed processes heats the lunar materials to such a high temperature that the iron oxide decomposes to liberate oxygen gas. Another dissolves ilmenite or some other iron mineral in sulfuric acid to release oxygen. Another uses hydrofluoric acid to release oxygen from unprocessed lunar regolith. Yet another uses fluorine as the reactant. Another uses a cold plasma of chlorine atoms to displace oxygen from minerals. (Chlorine atoms, made by shining raw sunlight on chlorine gas, react with a metal oxide to make oxygen plus a metal chloride. The metal chloride is electrically decomposed, regenerating chlorine and providing the metal as a bonus.) Many of these processes are promising but have not yet been adequately studied because of the chronic shortage of funding for laboratory research on the uses of non-terrestrial materials.

Schemes for extracting oxygen from lunar rocks usually feature melting or chemically destroying the principal minerals in the rocks and regolith. They therefore are also very efficient at releasing all the trapped solar-wind gases in the lunar material. Those schemes that concentrate on a single mineral instead of unsorted raw material almost always choose ilmenite as the target, because it contains the lion's share of the solar-wind gases. Therefore, all of the processes envisioned for extraction of oxygen will release solar hydrogen, helium, and other gases as byproducts. Although, as we have seen, the hydrogen content of lunar materials is small, it may be sufficient to make up for hydrogen losses incurred in the production of oxygen. Some hydrogen will become trapped in the waste materials (iron, titanium dioxide, dross, and unreacted ilmenite) from oxygen production, and some hydrogen will diffuse through the walls of the reaction vessel or leak past valves and escape into space. There is reason to hope that the process can be designed so that the release of solar-wind hydrogen will fully compensate for these losses. The oxygen extraction process may then become wholly independent of resupply from Earth. This source of hydrogen is not, however, a credible source of rocket propellant for use in the return to Earth because the amount of hydrogen produced, even if it is larger than that required for makeup of losses, is still thousands of times smaller than the amount of oxygen.

When a spacecraft returns from the Moon, it may either land directly on Earth or return to the space station. Any vehicle that returns to Earth's surface must be replaced by the launching of a replacement vehicle, at great expense. But a vehicle that returns to the space station may be sent on another mission to the Moon as soon as it is refueled. Only the propellant for the first outbound trip to the Moon (plus a small mass of hydrogen for the return trip) must be launched from Earth. Thus using the space station as a transportation hub significantly reduces the mass that must be launched from Earth.

The vehicle returning from the Moon can get to the space station for unloading and refueling in either of two ways. The most obvious is that the vehicle blasts off from the Moon, kills a portion of its orbital velocity around Earth, and drops in on a highly eccentric orbit that lies in the plane of the space station orbit and reaches from the Moon's orbit down to the space station's own altitude above Earth. Three days later, when the lunar vehicle has fallen all the way down to the altitude of the space station, it fires a rocket motor to kill part of its velocity and drop it into a nearly circular orbit close to the space station's own orbit. It is then a simple and leisurely matter to rendezvous and dock with the space station.

The second method of returning from the Moon to the space station involves a nearly identical launch from the Moon. The spacecraft again enters a highly elliptical orbit about Earth, except that in this version its point of closest approach to Earth is not at the altitude of the space station (probably 500 to 600 kilometers up), but at an altitude of only 80 to 100 kilometers, actually within Earth's upper atmosphere. There the vehicle, protected by a lightweight heat shield, passes through the upper atmosphere and is slowed sufficiently to drop the apogee (high point) of its orbit down to approximately the altitude of the space station. The vehicle then coasts up to its new apogee of about 600 kilometers and briefly fires a small rocket engine to give it just enough extra speed to lift its perigee (low point) out of the atmosphere. The vehicle can then rendezvous and dock with the space station with little further expenditure of fuel.

The first (all-propulsive) method uses much more propellant than the second. The fuel required to drop into the space station's orbit makes up about 50 percent of the mass of the returning lunar vehicle. The second (aerobraking) method, which needs only about 5 percent of its mass as fuel for rendezvous with the space station, also requires, and largely burns up, a heat shield with a mass of about 15 percent of the vehicle mass. The latter therefore has a significantly smaller take-off weight from the Moon and requires less lunar propellant. But it also requires a heat shield to protect it during aerobraking. If the only source of heat shields is Earth, then every mission returning to the Moon from the space station would require not only refueling with propellants from Earth but also a new replacement heat shield lifted from the ground. The fuel supply needed to reach and land on the Moon must be increased because of the additional weight of a heat shield that must be carried from the space station to the lunar base. This greatly reduces the benefits of using aerobraking. It is natural to ask at this point whether it is possible to make the aerobrake heat shield on the Moon rather than on Earth. If so, we could realize the benefits of aerobraking without paying the penalty of lifting the heat shield from Earth to the space station, lifting it from the space station to the vicinity of the Moon, and landing it safely on the Moon.

Heat shields are made of materials, called refractories, that are very difficult to melt and vaporize. Refractories usually have very high concentrations of one or more of the oxides that are least volatile, such as calcium oxide (lime), aluminum oxide (alumina), or titanium oxide (rutile). But it happens that one of the two major byproducts of lunar oxygen production from ilmenite is rutile. Research by Professor Tom Meek, of the University of Tennessee, has shown how to weld loose rutile grains into a very strong heat shield. The method of choice is to heat the bed of rutile grains in a mold in what is effectively a high-power microwave oven. The grains then soften at their points of contact and fuse together, a process called sintering. When cooled, these microwave-sintered rutile masses are much stronger and resistant to thermal shock than conventionally heated rutile. Thus, it appears feasible to manufacture practical heat shields in reusable molds on the Moon. The ilmenite reduction reaction automatically provides 5.4 tons of rutile for each ton of oxygen extracted, of which only about 1.6 tons are needed for each mission that uses 9 tons of propellants (1.6 tons rutile per 8 tons lunar oxygen, or 0.2 tons rutile per ton of oxygen, only a low percentage of the rutile production rate). If this is done, then two of the three products of ilmenite reduction will be in routine use. Both rutile and iron will be accumulating in stockpiles on the Moon.

The second byproduct of ilmenite reduction is high-purity metallic iron, which can be directly melted and cast into products. Alternatively, the iron can be volatilized by reacting it with carbon monoxide gas at a temperature near 120 degrees Celsius to make gaseous iron carbonyl. This iron carbonyl can be further purified by distillation and decomposed by moderate heat (around 200 degrees Celsius) to deposit ultra-high-purity iron, with complete recovery of the carbon monoxide gas. Each ton of oxygen extracted from ilmenite makes 3.5 tons of iron metal available. Extremely strong and as corrosion-resistant as stainless steel, ultrapure iron makes a superb structural material for building propellant tanks for rockets or new lunar-base modules. The extraction of enough oxygen for a single return trip to the space station (eight tons) makes twenty-eight tons of iron, enough to build a new lunar-base module. These modules are by far the most massive part of the lunar base. Thus, once ilmenite reduction is in use on the Moon, the use of metallic iron as a structural material can also begin, and reliance on Earth as a supply source can be scaled down dramatically.

To summarize, a relatively simple complement of equipment to beneficiate ilmenite ore, electrolyze carbon dioxide or water, sinter rutile, and carbonyl-extract iron permits an early lunar base to meet a large proportion of its own material needs, thereby substantially defraying the expenses of construction and operation. The ability to raise a major fraction of its own food would further increase the base's material and economic independence.

Many other uses of lunar material may become feasible after the installation of a sizable lunar base. Glasses and ceramics extracted from the regolith may be used as structural materials, paving blocks, or windows. Thin-film amorphous-silicon solar cells may be manufactured to expand the electrical power supply to the base. The combination of lightweight solar cells with power management (production and storage) using a reversible fuel cell may actually turn out to be superior in performance to a nuclear reactor. Such an option would be far more acceptable to environmentalists than launching even an inert reactor from Earth.

The astute reader has probably noticed by now that every single use of lunar resources described in this chapter serves to increase the autonomy of a lunar base and lower its operating costs. There has been no mention of any way to *profit* from the use of lunar material. The reason for that is simple: a new lunar base is like a new baby, just beginning to understand and develop its powers. No one expects a newborn baby to go out and get a job before learning the basic life skills and getting schooling. Like a new lunar base, a baby must first master at least the arts of breathing and eating. But of course we all know that most babies do eventually turn into self-sufficient adults, whereas we know nothing about whether a lunar base will ever be able to pay its own way. It is true that the base could still have a very bright future even it were never completely self-sufficient in all commodities—this is, after all, just how individuals, communities and nations on Earth function. What the base needs is something to export that will generate enough revenue to pay for the things it needs but cannot produce on its own. It needs to have a paying job.

What might a lunar base export for income? Scientific samples will be in demand, but, given the enormous sensitivity of modern analytical techniques, it seems most improbable that enough mass of samples would be wanted to generate a significant income stream. In the relatively near term, only two things are valuable enough per ton that their export from the Moon might sustain the lunar economy: rocket propellant for use in support of the logistics of other large-scale operations originating in near-Earth space and surplus energy for export to Earth. Both of these topics lie in a future somewhat beyond the early lunar-base era covered in this chapter. Both deserve, and will receive, more careful scrutiny.

5

ASTEROIDS AND COMETS
IN OUR BACKYARD

"Ah, Houston, Ares One here. We're suited up and closing with our target at eight meters per second, range just under ten klicks. We're showing less than one meter per second crossrange. We're go for landing. Beginning rotation maneuver to put our legs out front." The star field, with the great irregular bulk of the asteroid Scotti in the center, shifted rapidly to the side. The asteroid vanished from the front windows. "Ah, OK, rotation complete. Hydrazine thruster tests in progress. Here's +Z, −Z, +Y, −Y, +X, −X. All check out okay. We'll be back on voice for the landing."

Wiz thumbed off the transmitter and went to intercom. If there was an answer, it would take a half hour to get back to them. Let the computer listen; they had work to do.

Wally came up on the intercom from the instrument module, the one they would call the Mars Science Center on the next trip out. "Hey, Wiz, how does it feel to be stuck at your desk while I get to take the first stroll on Scotti?"

"Some guys get all the luck." Wiz was playing it light, but he had to be hurting. Everybody knows how it feels for a real astronaut, a red-hot rocket jockey, to sit by while a crewmate makes history. They even have a name for it: the Collins Syndrome.

"Giving up so easy?" Wally was intent on playing it for all it was worth. "That's not like you, ol' buddy. This is mankind's second landfall. It's big!"

"I'm a busy man up here, Wally, got a timeline to run. Besides, I'm holding out for a really big one; you know, the big red one. Come on, read me through the checklist."

The cratered, shattered, dust-splotched visage of the near-Earth asteroid 7271 Scotti filled the screens of the aft monitors. A flashing magenta X on the display marked the landing site, just below center of the screen in the flat bottom of a kilometer-sized crater.

"OK, we're at minus 1:30 till thruster burn to null out our approach and crossrange velocities at a height of one klick above target. Then we're on gyro-lock until touchdown, free fall all the way to the surface unless a last-minute correction is needed. Latest comp estimate of the grav field says a free-fall landing speed of 1.4 meters per second. Walking speed. So no thruster burns except attitude control from here on in. Don't want to mess up that virgin asteroid with nasty Earth gases if we can help it."

"Yeah, gotta keep your credibility with the Sierra Club."

"OK, Wally, I need you strapped down on the instrument deck until landing. Gotta get those instruments up to speed. I'm already seeing GRS data up here, but we need the mags on pronto."

"I'm already on the spot and buckled in. My prelanding checklist says GRS active and feeding to comp, radar altimeter on-line and slaved to guidance, IRIS is off, mags on and in self-calibration cycle. OK, I just got green on the mags. Rear camera systems recording and feeding the main com antenna for the folks back home. The board is green, checklist completed.

"Internal hatches sealed in preparation for landing." Wally heard several distinct clunks as the latches dogged down. The thrusters fired briefly to null out their motion relative to the asteroid. "Please make sure your tray tables and seat backs are in their full upright and locked positions." Wally rolled his eyes.

"Looking good, Wally. We're at five klicks and closing on auto. You just sit tight now until landing. No more trips to the lavatory. Use the burp bag in the seat back in front of you."

A few seconds pause, and Wiz came back on. "OK, Wally, as soon as we're down and I have a confirmed green board, you're clear to head out the back door. Remember it's the camera on the +X leg that's on com to Earth, so don't go down the wrong leg." Wally glanced up at the +X screen and visualized himself in dignified descent to the strains of patriotic music. Very fine. "Now, let's keep the chatter down until we're landed."

"Roger that, Wiz," Wally muttered, recalling Wiz's seemingly endless gift of gab.

Scotti swelled in the rear-facing monitors. Wally really had nothing to do but watch. If there were an anomaly on the board, the computer would speak to him about it. Individual rocks were clearly visible on the +X screen, casting long, intense shadows. A chain of about two dozen tiny craters marked a crack in the underlying bedrock. Wally frowned at the screen for a moment. Something dark and blurry appeared at the edge of the screen. He reached for the com button but hesitated in confusion as the dark blur grew larger, hiding a third of the face of Scotti.

The hackles rose on the back of Wally's neck. "What in holy . . . what is . . ." He punched the com button, "Wiz!" A sudden flash of comprehension locked his jaw and fists. "What are you doing?"

Wiz's voice came back with the faint hum of suit-to-suit radio. "See you on the surface, Wally. I figure you'll be down about a minute behind me. Not bad, ol' buddy. You'll go down in the history books as the second man ever to land on an asteroid. By the way, I noticed the flight deck escape hatch is open. Better not go in there without your helmet on."

The amoeboid black blur emerged into full sunlight as it cleared the end of the +X landing leg, coming into better focus as it receded from the camera lens. No doubt about it, it was Wiz's suit. The blue hash marks on the legs and arms glowed like sapphires in the sunlight. Wiz waved broadly at the camera and the watching billions, then turned lazily to aim his feet directly at the landing area, drifting like so much thistledown.

"So, Wally, I guess you're in charge now. Don't forget the post-landing checklist. Good luck!"

By the time the TV signals got back to Earth it was far too late for the brass to intervene. The best they could do was put the audio on a delay loop and censor Wally's colorful commentary.

NASA annals report that both landings were uneventful.

—One Really, Really Big Step for a Man

The nineteenth century began with the discovery of the existence of asteroids. The first and largest asteroid discovered, Ceres, was found on the night of January 1, 1801. The asteroids were numbered in their order of discovery, starting with 1 Ceres, 2 Juno, 3 Pallas, 4 Vesta, and so on in an ever-accelerating series. Improvements in telescopes and the introduction of astronomical photography sped up the process. During the remainder of the century, over four hundred asteroids were discovered, almost all lying between 2.2 and 3.3 astronomical units (AU) from the Sun, taking 3.3 to 6.0 years to complete their orbit about the Sun. (The AU, the basic yardstick of the solar system, is Earth's mean distance from the Sun, about 150,000,000 kilometers, or 93,000,000 miles.)

Only in the closing months of the century, in 1898, was a fundamentally different type of asteroid discovered. The asteroid 433 Eros was found to follow an unprecedented orbit—orbiting the Sun every 1.75 years and ranging from 1.13 to 1.78 AU from the Sun. Earth's greatest distance from the Sun (its aphelion distance) is 1.017 AU, and Mars's mean distance from the Sun is 1.5 AU. Therefore, Eros in its present orbit can cross the orbit of Mars and pass distressingly close to Earth. Other discoveries were soon to follow. The asteroid 887 Alinda, discovered in 1916, also approaches to within 1.15 AU of the Sun. Discovered in 1929, 1627 Ivar, approaches to within 1.12 AU of the Sun, and 1221 Amor, found in 1932, approaches to within 1.08 AU of the Sun. The Earth-grazing asteroids that approach within 1.3 AU of the Sun but do not currently cross Earth's orbit are called the Amor family. All of the Amor asteroids cross the orbit of Mars and could collide with it. They also are subject to fairly strong and frequent gravitational perturbations by Mars, so that the size and shape of their orbits are subject to significant changes.

In addition to Amor, the year 1932 also saw the discovery of 1862 Apollo. Apollo's orbit takes it as far from the Sun as 2.29 AU and as close as 0.65 AU. Apollo therefore crosses not only Mars's orbit but also Earth's and Venus's orbits. The asteroids discovered since Apollo that have similar Earth-crossing orbits are collectively referred to as the Apollo asteroids.

In 1976 a small asteroid, 2062 Aten, was discovered in an orbit that actually circles the Sun faster than Earth. Aten approaches to within 0.79 AU of the Sun at perihelion and retreats only as far as 1.14 AU at aphelion, and therefore crosses the orbit of only one planet, Earth. Its orbital period is 11.5 months, only slightly less than Earth's year. Other Earth-crossing asteroids with orbital periods less than one year are referred to as the Aten group. The Atens and Apollos, all of which actually cross Earth's orbit, are called Earth-crossing asteroids. Also, some Amors are in orbits that from time to time drift in and cross Earth, and these asteroids are called episodic Earth-crossers. The Amor, Apollo, and Aten asteroids together are sometimes called the AAA asteroids or, more commonly, the near-Earth asteroids (NEAs).

The Earth-crossing asteroids cannot have resided in such orbits for the entire age of the solar system. A typical Earth-crosser braving the heavy traffic of the inner solar system can expect to survive only some tens of millions of years before suffering a fatal accident. The most likely fate is to collide with one of the terrestrial planets. Of these, the most likely single target is Earth, followed by Venus, Mars, Mercury, and the Moon, in that order. A significant fraction evade collision by being perturbed by Earth or Venus into orbits that cross Jupiter's orbit. Jupiter, with more than one hundred times the mass of all the terrestrial planets combined, plays the role of a cannonball thrown into a quiet game of billiards. Jupiter's immensely powerful gravity quickly fires these unfortunate asteroids out of the solar system, swallows them up, or even dumps them into the Sun.

By 1984 the number of known NEAs had grown to sixty-four. A systematic photographic search program initiated by astronomers Gene Shoemaker and Eleanor Helin was by that time accounting for several new NEA discoveries per year. In the dozen years since that time, several other groups of observers inspired by the success of the Helin-Shoemaker team have entered the field, and the discovery rate has accelerated dramatically.

The most innovative and exciting of the new search efforts is the Spacewatch program run by Professor Tom Gehrels at the University of Arizona. Spacewatch features a 0.9-meter (thirty-six-inch) dedicated telescope with a sensitive 2000 × 2000 pixel CCD (charge-coupled device) detector array, controlled by a computer. Operating from a dome on Kitt Peak, Arizona, the Spacewatch team members have achieved numerous firsts. They have repeatedly set new records for the smallest and nearest asteroids ever discovered, and they are now finding more new near-Earth asteroids than all the other observing teams in the world combined. The number of known NEAs of all sizes is now nearing four hundred.

Spacewatch has also found a surprising number of asteroids that not only cross Earth's orbit but follow nearly circular orbits with periods very close to one year. This discovery is a challenge to theorists who wish to explain where NEAs come from, and a concern to those who wish to predict NEAs' future motions. The reason for concern is simple: many of these slow-moving, nearby objects are exceptionally susceptible to Earth's gravitational attraction. With expected orbital lifetimes as short as ten thousand years, some are a hundred to a thousand times more likely to strike Earth in a given time interval than a normal Apollo asteroid. If these bodies have such short lifetimes, why do we see so many? How are they replaced? Rocks ejected from large impacts on the Moon could easily get into such orbits, but many of the asteroids in this group are far too large to have survived ejection from the Moon in one piece. Their origin is at present not understood.

Whatever their source, the discovery of this class of bodies provides us with a new and previously unexpected source of massive bodies that can encounter Earth at very low speeds and impact at just above Earth's escape velocity. As I discussed in detail in my 1996 book *Rain of Iron and Ice*, such slow-moving bodies present an enhanced threat to Earth's surface because they are more likely to penetrate the atmosphere and impact the surface with their energy nearly intact. These bodies with very Earth-like orbits are also extremely easy to reach for spacecraft missions launched from Earth. Interestingly, the most dangerous are easier to reach and land on than the Moon. We return to this subject in chapter 7.

In essence, Spacewatch and other search programs have found that Earth is embedded in an immense swarm of asteroids and comets that threaten us with collision and global mass destruction. The size of the total population can be estimated in three independent ways: from the percentage of sky covered in searches, from the percentage of accidental rediscoveries among the total NEAs found in searches, and from the cratering rate on the terrestrial planets. We find from all three sources that about 2,000 asteroids larger than one kilometer in diameter cross Earth's orbit, of which over 200 have already been discovered, up from about 50 a decade ago.

The typical Amor asteroid follows an elliptical orbit that ranges from about 1.1 or 1.3 AU at perihelion to 2.5 or 3 AU from the Sun at aphelion. The large majority of all Amors reach out into the heart of the asteroid belt near aphelion. The orbital period of the average Amor is about three or four years. In that time the Amor crosses the orbit of Mars twice, once inbound and once outbound on its eccentric path about the Sun. The orbital planes of the Amor asteroids are mostly inclined by less than twenty degrees to the central plane of the solar system, the ecliptic, in which the planets move.

All Apollo asteroids cross Earth's orbit; that is, their perihelion distance is less than Earth's aphelion distance from the Sun (1.017 AU). The typical Apollo asteroid may have a perihelion anywhere from 1.017 AU in to well inside Mercury's orbit near 0.4 AU. The aphelia of Apollos are mostly in the belt, although some range as far out as the orbit of Jupiter. The orbital periods are again mostly about three to five years. An Apollo may make as many as ten crossings of planetary orbits on each trip around the Sun. The orbital inclinations are mostly between zero and twenty degrees. The bodies with exceptionally circular, very Earth-like orbits constitute a small subset of the Apollos. Gehrels has suggested that these bodies be designated as a new NEA family, which he proposes to call the Arjunas. Many of these bodies have orbital inclinations of twenty to thirty degrees.

The Aten asteroids all have orbital periods less than one year and therefore have mean distances from the Sun less than 1 AU. They all have aphelia slightly outside Earth's orbit (they must in order to appear in the night sky and, hence, be discovered). It is reasonable to wonder whether there are swarms of small asteroids that orbit entirely within 1 AU of the Sun, which evade detection because they never cross Earth's orbit, there to appear in the dark night sky and catch the attention of insomniac astronomers. The existence of minor belts of asteroids between the orbits of the inner planets has sometimes been discussed, but we remain ignorant regarding their reality.

There is one final class of asteroids that may exist in the inner planetary system but that defies attempts at discovery. These are asteroids that are trapped in one-to-one resonance with a terrestrial planet (in which the orbital periods of the asteroid and planet are equal). The classical example of such behavior is the clustering of asteroids in two regions lying on Jupiter's orbit about the Sun. One of these clusters is centered on a point equidistant from Jupiter and the Sun, sixty degrees ahead of Jupiter on its orbit. The other is in the mirror-image point, sixty degrees behind Jupiter. They all have the same mean distance from the Sun as Jupiter (5.2 AU), the same orbital period (twelve years), and modest orbital inclinations and eccentricities.

When the first of these bodies were discovered in 1906, they were given the names of leading Trojan and Greek figures in the Trojan War, as described in Homer's *Iliad:* Hector, Patrocolus, Achilles, and so on. Several hundred such asteroids, referred to as the Trojan asteroids, are now known. The eminent French mathematician Joseph-Louis Lagrange was able to show (about 1780) that there are five special points in a gravitating system dominated by two large bodies (the Sun and a planet) which have unusual orbital properties. These are called the Lagrange points after him. The fourth and fifth of these points, designated L4 and L5, are stable points sixty degrees ahead of and behind the planet where Jupiter's Trojan asteroids are. The other three points are metastable, meaning that a body placed exactly at one of them would remain there, but an infinitesimal disturbance would cause it to drift away.

In principle, other planets besides Jupiter should be able to collect and retain Trojan asteroids on their own orbits. In 1990 the asteroid 1990 MB was found to have an orbital period exactly equal to that of Mars. Surprisingly, it follows a Lagrange point on Mars's orbit. Of all the terrestrial planets, the ones best able to retain Trojan asteroids should be Earth and Venus, not little Mars. The problem is that searching Earth's Lagrange points is extremely difficult. The L4 and L5 points on Earth's orbit are only sixty degrees away from the Sun as seen from Earth, and therefore can be seen in a dark sky only when very close to the horizon. Several searches for Earth Trojan asteroids have so far turned up no such bodies, but the discovery of a Mars Trojan tells us that the chances are good that there are some Earth Trojans as well. We may expect searches to continue with ever-improving instrumentation and techniques.

We have mentioned that about two thousand asteroids larger than one kilometer in diameter cross Earth's orbit. But vastly greater numbers of smaller bodies accompany the larger and more easily discovered ones. Before the debut of the very sensitive Spacewatch program, the practical size limit for photographic detection of NEAs was at a diameter of about 200 meters. In its first months of operation, Spacewatch demonstrated the ability to discover 100-meter, 50-meter, 20-meter, and even 10-meter asteroids. To date, several bodies smaller than 10 meters have been discovered, down to the present record of 4 meters. According to Spacewatch discovery statistics, the number of bodies with diameters larger than 100 meters (330 feet) that follow NEA orbits is about five hundred thousand. Even more astonishing, the number of 10-meter bodies in Earth-crossing orbits is roughly a hundred million!

A 10-meter asteroid, about the size of a typical single-family residence, does not sound like a very substantial body. It would weigh in at about one thousand tons, about 1 percent of the mass of a large nuclear-powered aircraft carrier. However, because of its orbital motion and acceleration by Earth's gravity, the typical Earth crosser would impact Earth at a speed of about twenty kilometers per second, carrying a kinetic energy equivalent to the explosion of 100 kilotons of TNT. For comparison, the atomic bomb explosions that devastated Hiroshima and Nagasaki were about 20 kilotons each.

Since the lifetime of the average Earth-crosser is about 30 million years, the 100 million ten-meter asteroids are being lost at an average rate of three per year. Of these, according to orbital evolution calculations, about a third are lost through collision with Earth. Thus, we should expect, on average, about one 100-kiloton explosion each year caused by Earth impact of a small asteroid. A significant fraction of all these impacting bodies are so physically weak that they break up at high altitudes, dissipating their enormous energy where it does little or no damage to people and property. The physical strength of these bodies is a major factor in determining how hazardous they are. Unfortunately, determining the physical properties of an asteroid before its impact is quite challenging.

The principal source of information on the physical properties of asteroids has long been the measurement of the changing brightness of the asteroid over the course of its rotation. The measured brightness variations over the rotation period are called the light curve of the asteroid. Light curves may show varying brightness due to two independent factors: local variations in reflectivity and departures from spherical shape. A perfectly spherical asteroid with perfectly uniform surface reflectivity (*albedo*) would show constant brightness. We therefore would not even be able to determine its rotation period. An elongated body would display two brightness peaks per revolution, corresponding to the two side views. A body with large randomly distributed bright and dark spots would show a complex variation of brightness with time during each rotation, but the entire pattern would repeat itself faithfully during the next rotation.

How could we determine whether an asteroid with a nice smooth, reproducible three-hour cycle of brightness variations is a spotted sphere rotating every three hours or an elongated ellipsoid rotating every six hours? There is in fact a way to do this even for bodies so small or so distant that they look like a point of light. The principle is rather simple: If the body varies in albedo as it rotates, then the darker areas will be hotter than the light ones. All of the sunlight that strikes the body is either reflected or absorbed. Absorbed sunlight heats the surface. The hotter the surface gets, the more infrared (heat) radiation it emits. The brighter areas, which reflect more and absorb less, are cooler and give off less infrared. Dark areas absorb more, are hotter, and give off more infrared radiation. If the body is spherical with varying albedo from place to place, then the *total* amount of visible and infrared energy coming from its surface is constant: when one goes down, the other must go up by the same amount. On the other hand, if the body is uniform in albedo and elongated in shape, the temperature will be the same everywhere, but the apparent cross-section area of the body will vary widely as it revolves. The visible brightness and the infrared brightness will vary in unison, and the total will vary markedly with time. On a real asteroid that is both "blotchy" in color and irregular in shape, simultaneous measurement of the visible and infrared brightness of the body throughout a rotation permits us to separate these two effects.

It has been known for many years that the asteroid Eros is markedly elongated, so that the visible cross-section area may change by a factor of four as it rotates. The light curve of the NEA 1620 Geographos is even more extreme, with the apparent cross-sectional area varying by about a factor of nine with each rotation. If this variation is due to shape alone, Geographos must have more the shape of a pencil than a ball or an Idaho potato.

In recent years a powerful new radar technique for determining the shapes of NEAs has become available. Steve Ostro of the Jet Propulsion Laboratory and his colleagues have constructed radar images of several NEAs during unusually close approaches to Earth. They have found an astonishing tendency for NEAs not just to be irregular or elongated but to be made of two large lumps in contact with each other, a bit like a cartoonist's version of an old-fashioned dumbbell. Such a shape is called a contact binary. The most likely origin for a contact binary is incomplete fragmentation of a somewhat larger asteroid in a catastrophic collision, spraying small pieces of debris about and leaving two large pieces still gravitationally bound to each other, orbiting about their center of mass. The gravitational fields of these irregular chunks are so lumpy that any orbit is certain to evolve rapidly. In a short time, these pieces will wander into collision, dissipating part of their orbital energy through the crushing of more rock, and leaving them rotating at a speed close to orbital velocity in contact with each other's equators.

Picture yourself on the surface of a rapidly rotating dumbbell-shaped asteroid. (I will not hazard a name for this body: many asteroids are named after prominent astronomers, but no one has had the temerity to attach the name of any particular person to a dumbbell.) Suppose you land at one end of the dumbbell. The surface gravity there is very weak because the normal small gravitational attraction is almost canceled out by the centrifugal force of its rapid rotation. The horizon will be strangely close because of the small radius of the body, only a few kilometers. As you walk toward the "north pole" the strength of the gravitational force increases, but it no longer pulls you in the direction your eyes tell you is "down": it pulls you instead toward the point of contact of the two bodies, which is approximately where the center of mass of the asteroid is. Your impression will be that you are descending an ever-steeper slope. When you get about a quarter of the way around the circumference of the spheroidal lobe on which you landed, the other lobe will suddenly appear looming above the horizon in front of you. You will see it as a great mound becoming visible at the bottom of a steep slope that you are descending. As you descend the slope toward the contact point of the two lobes, the acceleration of gravity gets steadily weaker and the direction of the gravitational force gets ever closer to horizontal. You also notice that, as you descend the surface, your path constantly deviates to the left of the direction in which you are heading. After a while the force is so close to horizontal that you begin to slide in slow motion down the slope toward—and to the left of—the contact point. The second lobe looms overhead until it fills most of the sky. You have to hold on to the rocks as you descend to keep from being diverted away from your goal by the strong Coriolis (gyroscopic) force caused by the asteroid's rapid rotation. In free fall, but grasping every rock available to control your descent, you arrive at the contact point, where the two lobes lie crushed against each other. You see a broad band of crushed, flattened rock girdling each lobe, where the two halves of the dumbbell first contacted each other and rolled along each other's surfaces until friction brought them to a standstill. You are now nearly weightless, at the bottom of a narrow, steep-walled valley, precisely at the asteroid's north pole. As you make your way along the valley, you quickly circumnavigate the constricted waist of the asteroid, finding yourself back at the point where you first descended into the contact zone. In doing so, you pass over the south pole and return to the north pole within a distance of perhaps a hundred meters. By this point you have probably decided to settle somewhere else.

The continuous bombardment of asteroid surfaces by small impactors generates a constantly replenished supply of crushed rock. The larger the asteroid, the more of this cratering ejecta is captured by the asteroid's gravity. Large asteroids should have deep regoliths of impact-shattered rock and glass formed by impacts. The thickness of the regolith should be progressively less for smaller asteroids, which lack sufficient gravity to capture their own cratering debris. The NEAs less than one kilometer in diameter probably are incompletely covered by thin layers of regolith. Our direct evidence about regolith thickness on asteroids is very limited. The two asteroids studied by the *Galileo* spacecraft en route to Jupiter, Gaspra and Ida, are both much larger than the NEAs of present interest to us, and both are, as expected, essentially completely covered by regolith.

The same impact processes that make roughly equal-sized contact binaries and regolith also can leave small collision fragments in high enough orbits that they can persist for many millions of years. Earth-based astronomical studies have given several hints of small satellites in orbit about large asteroids, but the clinching evidence came in photographs taken by the *Galileo* spacecraft that clearly show a kilometer-sized moon, christened Dactyl, in orbit about Ida. Study of Dactyl to date suggests that it is covered by regolith.

The fragmentation behavior of an asteroid depends on how strong the asteroid is. A fluffy dustball and a giant crystal of stainless steel will behave very differently during a high-velocity collision. It is clearly important to find out what asteroids are made of.

One useful source of information on the composition of asteroids has been studies of their reflection spectra, much like the studies used to produce mineral maps of the Moon. The ultraviolet, visible, and near-infrared spectra of several hundred asteroids have been measured for this purpose. These spectra have been compared to laboratory spectra of pure minerals, terrestrial and lunar rocks, and samples of many different types of meteorites.

In general, the large majority of the asteroids have spectra that resemble either known classes of meteorites, or mixtures of familiar meteorite minerals in somewhat different proportions than are found in actual meteorites. The properties of meteorites tell us an enormous amount about the properties of asteroids.

A meteorite is any object that falls from space and survives to land on Earth's surface. Any particle that burns up in the atmosphere and does not survive to reach the ground is called a meteor. The Sun-orbiting pieces that are potential meteors and meteorites are called meteoroids while they are still in space, before encountering Earth's atmosphere. Since the objects that burn up (meteors) tend to be physically weaker than those that penetrate to the ground, our compositional information on meteorites must be heavily biased in favor of the strongest interplanetary meteoroidal and asteroidal materials.

Over ten thousand meteorites have been collected on Earth's surface. These meteorites are divided into two groups; those that were observed to fall (falls), and those that were not observed to fall but were simply found on the ground (finds). There are large differences in the meteorites found in these two categories. Two of these differences serve to illustrate the principle. First, the very strongest meteorite type, metallic iron, makes up only about 3 percent of the falls, but at least 30 percent of the finds. Second, the most fragile meteorites, the CI (Ivuna-type) carbonaceous chondrites (see the following paragraphs) are known only among falls. The reason for these differences is straightforward: meteorites that are very resistant to crushing and weathering survive better on Earth's surface and show up among the finds with disproportionate frequency, whereas those weakest and most vulnerable to weathering literally may not survive a single rain shower and must be seen falling to have any chance of recovery. The statistics of observed falls are therefore much more useful guides to the true relative abundance of meteoroid types in space than the statistics on finds, but even these are strongly biased against weak materials that break up and burn up during entry into the atmosphere.

Since the first acceptance of meteorites as extraterrestrial material in the early 1800s, three major classes of meteorites have been recognized. These are irons, which are typically 99 percent iron-nickel-cobalt metal alloy with about 1 percent of assorted other minerals; stony-irons, which contain about 50 percent metal and 50 percent silicates; and stones, which are dominated by silicates and contain less than 31 percent (and usually 0 percent to 20 percent) metal, usually with a small percentage of iron sulfide.

Modern studies of meteorites, informed by our current understanding of the processes of melting, separation, and crystallization of rock types (the science of petrology), distinguish approximately fifty chemically distinct classes of meteorites. The most useful classes to understand are those that are most common or that have reflection spectra most closely resembling those of asteroids.

By far the most common types (87 percent of all falls) are stony meteorites containing tiny (usually 0.1 to 1 millimeter) glassy or partially glassy nearly spherical particles called chondrules, after the Greek word for "seed" or "droplet." These meteorites, called chondrites, are composed principally of silicate minerals, usually with some metallic iron and sulfides. The relative abundances of the rock-forming elements in the chondrites are closely similar to their relative abundances in the Sun and in cosmic material in general. They show no signs of separation into materials of different density as would occur during partial melting in a gravity field, and their textures attest to the absence of melting since they were originally accreted from their component particles. Their ages generally cluster very close to 4.55 billion years, a date that from several other lines of evidence must correspond closely to the time of origin of the solar system. Chondritic material is the most ancient, least processed material in our planetary system, a sample of the raw materials available at the time of the accumulation of the planets.

The chondrites fall naturally into five composition classes, of which three have very similar mineral contents, but different proportions of metal and silicates. All three contain abundant iron in three different forms (ferrous iron oxide in silicates, metallic iron, and ferrous sulfide), usually with all three abundant enough to be classified as potential ores. All three contain feldspar (an aluminosilicate of calcium, sodium, and potassium), pyroxene (silicates with one silicon atom for each atom of magnesium, iron, or calcium), olivine (silicates with two iron or magnesium atoms per silicon atom), metallic iron, and iron sulfide (the mineral troilite). These three classes, referred to collectively as the *ordinary chondrites,* contain quite different amounts of metal. Those with high metal content, about 23 percent, are called the H (high-iron) chondrites. Those with about 16 percent are called the L (low-iron) chondrites, and the relatively small class with about 6 percent metal are called the LL (low-low iron) chondrites. The H and L chondrites are both very common. The H chondrites are most similar in composition to Earth.

The two other classes of chondrites differ dramatically from the ordinary chondrites in their mineral content and oxidation states. One of these is very strongly reduced, with virtually no oxidized iron in its silicates. The amount of metal is very large, reaching as high as

31 percent, coexisting with an abundance of the iron-free magnesium silicate $MgSiO_3$, enstatite. Minor or trace amounts of several very unusual sulfide and nitride minerals are also present in these E (enstatite) chondrites. The remainder of the chondrites are unusual for the opposite reason: they are the most strongly oxidized of meteorites, containing ferric iron, carbon, organic matter, and little or no metal. Because of the prevalence of soot and organic matter in them, they are referred to as the C (carbonaceous) chondrites. There is much idiosyncratic variation among the C chondrites, but four main subclasses are generally recognized. Two are rather similar to ordinary chondrites, containing the least ferric iron, the least organic matter and carbon, and some traces of metal. But the CM (Murchison-type) chondrites contain essentially no metal, about 10 percent water bound in hydrated silicates (clay minerals), up to 3 percent organic matter, elemental sulfur, and some carbonates. Finally, the most oxidized and most volatile-rich meteorites are the CI (Ivuna-type) chondrites, which contain up to 20 percent water in clays, up to 6 percent organic matter, and abundant carbonate and sulfate minerals. Ferric iron in the form of magnetite is abundant, and metal is completely absent.

About 7 percent of all falls are heavily recrystallized and fractionated stones with virtually undetectable traces of metal and sulfide, evidently separated according to density in a weak gravitational field. Since these stones never contain chondrules, they have been given the name *achondrites*. Most achondrites are very old, dating from immediately after the formation of chondrites, perhaps even overlapping that time. Some achondrites have densities and chemical compositions similar to those of planetary crustal rocks; others are made of denser rock with high magnesium and iron content, very similar to that of planetary mantle rocks. Indeed, several achondrites are typical Moon rocks ejected from the Moon by impact events, and four rare types of achondrite have exceptionally short ages (as little as 1.5 billion years) and contain occluded gases whose abundance pattern constitutes a fingerprint of the atmosphere of Mars.

Leaving aside these rare lunar and Martian meteorites, most achondrites are asteroidal in origin. There has been occasional speculation regarding a possible origin of another rare class, the enstatite achondrites, on the surface of Venus or Mercury. At the moment, the theory of ejection of meteorites from planetary impacts suggests that getting an ejecta fragment out of the dense Venus atmosphere is essentially impossible. Unfortunately, our knowledge of the surface of Mercury is still too rudimentary to tell whether any achondrites come from there.

About 3 percent of all falls are irons. Irons come in a variety of classes with different crystal structures, chemical compositions, and inclusions of other minerals. The three principal elements in irons are always iron, nickel, and cobalt. The nickel concentration can range from about 5 percent to 60 percent. The cobalt abundance is usually about a tenth of the nickel concentration. In general, irons represent recrystallized core-forming melts ejected by surface impacts. They are likely to be either core material from middle-sized asteroids that have been completely destroyed by impacts or "raisins" of metal formed in the interiors of asteroids that were so small they lacked the gravity to draw all the metal into a well-defined core.

Closely related to the irons in mode of origin are stony-irons, which make up about 2 percent of all falls. These are found in two distinct classes, the pallasites, which consist of a continuous metal matrix with embedded lumps of silicates, and mesosiderites, which are a mixture of lumps of metal and silicates. The pallasites are most clearly of igneous origin.

As noted earlier, the ability of an entering body to penetrate the atmosphere and survive deceleration depends on its crushing strength. The basic criterion for survival is that the crushing strength be greater than the maximum aerodynamic pressure experienced during atmospheric entry. The pressure, which is easily calculated by multiplying the local atmospheric density (grams per cubic centimeter) times the square of the entry velocity (centimeters per second), can be enormous. The minimum entry velocity is simply Earth's escape velocity, which is the speed that a body would have if it fell from a state of rest at great distance from Earth, accelerated all the way to impact by Earth's gravity. For Earth, the escape velocity is 11.2 kilometers per second (1.12×10^6 cm/s). The maximum possible impact velocity for a solar system body striking Earth is a little more complicated to calculate. The fastest a body can travel at Earth's orbit and still be a part of the solar system (that is, gravitationally bound to the Sun) is the local escape velocity of the Sun. That is equal to the local circular orbit velocity (that is, Earth's orbital velocity about the Sun, 30 kilometers per second) multiplied by the square root of 2, which is 1.414. If this body approaches Earth head-on, then the relative velocity due to their combined orbital velocities is 30 × (1 + 1.414), which is 72.4 kilometers per second. This calculation so far neglects the acceleration of the body due to Earth's gravity. The energy added by falling in Earth's gravity field is the escape energy, which is added to the energy the body has due to relative orbital motion to get the total impact energy. That energy converts back into a velocity of 73.26 kilometers per second. The velocity added by Earth's gravity to this already fast-moving projectile is so small because the acceleration by Earth's gravity acts on the body for only a short time as it streaks toward impact.

In our solar system, the only bodies that can pursue orbits about the Sun in a direction opposite to the motion of the planets are long-period comets, which have randomly distributed orbital inclination angles. Asteroidal fragments (meteorites) usually enter at speeds between 12 and 36 kilometers per second. A cometary fragment entering at 72 kilometers per second experiences an aerodynamic pressure thirty-six times as large as that felt by an asteroidal fragment traveling 12 kilometers per second at the same altitude.

The aerodynamic force is proportional to the local density of the atmosphere (the denser the air, the harder it is to move through it). Since the density of air increases almost exponentially with depth in the atmosphere, a body may readily survive passing through the rarefied upper atmosphere where the density is one thousand times less than at sea level, but still get into serious trouble when it has penetrated to within a few kilometers of the ground.

A body traveling 12 kilometers per second through the atmosphere at sea level experiences a dynamic pressure of about 1,400 atmospheres. This is a serious problem for many geological and asteroidal materials. Iron meteorites, the strongest natural projectiles, have crushing strengths of about 3,500 atmospheres. Pallasites are nearly the same strength, and many achondrites and mesosiderites are about half as strong. Most stony meteorites range from about 1,500 atmospheres down to about 100 atmospheres in crushing strength. The carbonaceous chondrites not only are weaker but also vary greatly in strength. The relatively hard CV and CO chondrites may have strengths up to 1,000 atmospheres; but the CM chondrites probably range from 10 to 100 atmospheres, and the CI chondrites may have strengths as low as 1 atmosphere. In general, the C chondrites can survive entry only if they enter at a low speed and at a very shallow entry angle, so that they have time to decelerate at altitudes of tens of kilometers while experiencing peak pressures that do not exceed their strength. Clearly, irons and pallasites have a problem only if they enter at speeds substantially above the minimum. Hard stony meteorites span the entire spectrum of strengths between these extremes, and they will be selected accordingly.

The best conclusions we can offer are that irons and the hardest stones mostly survive entry without serious discrimination based on strength, but weak (especially C chondrite) projectiles must be enormously more abundant in the bombarding flux than they are in our museum collections. Several meteorite-entry studies have suggested that C chondrites make up about 50 percent of the entering bodies, not the 1 percent or 2 percent that the fall statistics would suggest.

Comparison of laboratory studies of the spectra of meteorites with astronomical spectra of asteroids provides us with an independent means of determining what NEAs are made of. An elaborate system of spectral classes has been established by a persistent cadre of asteroid observers. They honed their skills on the large, slow-moving asteroids in the belt, and in recent years they have run their total of asteroids studied to well over six hundred. They need moderately bright targets for their spectral studies. Also, since the equipment used to discover asteroids is completely different from that used to determine their spectra, these two functions are usually performed by different research groups at different observatories on different nights (and often in different years!). But large telescopes are in great demand and are often booked up months in advance. Most of them are very faint, and many are so close that their angular rate of motion across the sky is several degrees per day. A newly discovered NEA may fade to inscrutability in a couple of days, as observers struggle in vain to get telescope time on already-booked telescopes and transport their observing equipment to the most obliging observatory. If through logistic or weather reasons they miss that particular flyby, it will usually be several years before that new NEA again passes close to Earth. But most of the new NEAs of interest to us have been discovered so recently that they have not yet had a chance to come around again! For these reasons, spectral data on the NEAs are relatively sparse. In the face of these difficulties, it is remarkable that the spectra of some forty-five NEAs have been studied.

The spectral classes established by asteroid observers are based on the colors and total reflectivity of their targets. Those asteroids that are very black and have nearly featureless spectra in the ultraviolet, visible, and near-infrared appear in all respects very similar to C chondrite meteorites; they are therefore called C asteroids. Several other dark spectral classes, including the D, P, B and G types, appear closely related to, but distinguishable from, the C types. They appear to be made of the same materials as the C asteroids, but in somewhat different proportions.

Asteroids with another distinctive type of spectrum, much higher in reflectivity and with a hint of an absorption feature near 0.9 micrometers wavelength, are clearly composed of the same minerals that dominate most stony meteorites, but with different relative abundances. A debate between two schools of meteoriticists has raged for several years regarding whether these stony (S-type) asteroids are more closely akin to achondrites or to ordinary chondrites. The dilemma was that, until 1995, not a single belt asteroid had been found with a spectrum matching that of the most common classes of meteorites that fall on Earth, the ordinary chondrites. But now that difficulty has been remedied, and the easier interpretation of the S asteroids remains: they are most likely achondrites.

The asteroid belt is dominated by the C and S classes. The inner half of the belt is mostly S type, and the outer belt shades from C into D and P types, all of which are apparently carbonaceous.

The inner edge of the belt, the part closest to Mars and the other terrestrial planets, is compositionally complex. One class of asteroids found in that region has the reflection spectrum of meteoritic metal and is called the M class. Another class has extremely high reflectivity with absolutely no evidence of absorption of light by ferrous iron. These are called the E asteroids after their apparent dominant mineral, enstatite, and the enstatite achondrite meteorites they most closely resemble.

The NEAs show a broad diversity of spectral types, spanning most of the major classes seen in the Belt. The dark spectral classes account for about 25 percent of the NEAs that have been studied. About 50 percent of the sample are S asteroids. But even these results are not without bias. Obviously, we cannot take the spectrum of an asteroid unless we first discover it. But all NEAs are discovered by searching for them in visible light, either photographically, or with an electronic detector such as a CCD. A typical C asteroid with a reflectivity of 4 percent in visible light is far less easily seen than an equal-sized, equally distant S asteroid with a reflectivity of 16 percent. Supposing we have a detection system that has the ability to find NEAs down to some particular brightness level. A C asteroid with that level of brightness must have four times the area, twice the radius, and eight times the volume of an S asteroid of equal brightness. Say we find three S asteroids of that brightness for every C type. The C type, however, has eight times the volume and five times the mass (C material is much less dense than S material) of a single S type, which means that 55 percent of the mass of the asteroids of that brightness is C-type material. By reasoning of this sort, asteroid experts have concluded that roughly half the material in Earth-crossing orbits is carbonaceous. This is similar to the fraction of C-type material in the asteroid belt itself. In addition, studies of the fragmentation and burnup of meteor fireballs in the atmosphere suggest that the majority of their material is weak and carbonaceous.

Among the forty-five NEAs whose spectra have been studied are two M-type asteroids (4 percent). This compares very well to the approximately 3 percent of irons among meteorite falls. Three small NEAs have spectra very similar to those of basaltic achondrites and of the asteroid Vesta. They are members of the V spectral class. The NEA population also contains several bodies with spectra similar to those of ordinary chondrites, and these asteroids are members of the Q class.

By combining meteorite and asteroid data, we have come to a good general understanding of the nature of the bodies that fly through near-Earth space. But this improved physical and chemical understanding of the NEAs leaves the problem of their origin unanswered.

Since the mean lifetime of NEAs is about thirty million years, we can only account for their presence by postulating some mechanism for their replenishment. One possible mechanism has asteroids wandering, as a result of small collisions and the effects of solar radiation pressure, into orbits harmonically related to Jupiter's orbital period. These resonant orbits are very unstable. Jupiter's periodic gravitational perturbations rapidly pump up the eccentricity of the asteroid's orbit, causing its perihelion distance to drop down into the terrestrial planet region. Theoretical studies of this mechanism suggest that it works well but cannot supply more than about half of the observed NEA population.

A valuable clue to the problem of resupply of asteroids came to scientists' attention in a strange way. The story began in 1949 with the discovery of Comet Wilson-Harrington. Photographs of the comet taken shortly after its discovery show a clear but faint tail. Unfortunately, the comet was lost shortly after discovery. The comet had been followed for such a small portion of its trip around the Sun that the orbit calculated for it was not very accurate. Searches for the comet in later years verified that the orbit was not very good because the comet did not appear when and where the rough calculations predicted. And so the story languished for many years.

In 1979 a new NEA was discovered and named 1979 VA. A tentative orbit was calculated and run back in time a few years to see if it could be linked to any earlier "orphan" observations of asteroids. No such linkage was found.

Finally, in 1992 another "new" NEA was discovered. After enough observations had accumulated over a decent orbital arc, the orbit was run back for several years to search for earlier "one-night stands" in which the body was seen and reported without sufficient data to calculate an orbit. The calculated orbit linked up nicely with 1979 VA, showing that they are the same body. But that provided an orbit with a time baseline of thirteen years, sufficient for a very accurate calculated orbit that was good enough to justify running it much farther back in time to search for earlier accidental sightings. Much to everyone's astonishment, the "asteroid" 1979 VA turned out to be identical with the "comet" Wilson-Harrington 1949! The body that had been photographed with a definite cometary tail in 1949 was now a bare rock. Astronomers had accidentally caught a comet in the act of turning into an asteroid.

Theorists had already looked at the evolution of comets and their orbits over time. They knew that accidental close passes of long-period comets by Jupiter or Saturn could kick them into typical short-period comet orbits, penetrating deep into the terrestrial planet region. A series of theorists, beginning with the astrophysicist David Brin (who was then barely launched on his much more satisfying career as an award-winning science-fiction writer), showed that a comet stuck for millions of years in a short-period orbit would gradually lose the ices in its outermost meter or so, leaving behind a very black, fluffy lag deposit of carbonaceous dust of the sort routinely seen in comet tails. This dust deposit, once it is some tens of centimeters thick, is so opaque to sunlight and such a poor conductor of heat that it effectively blocks the cometary ices in the interior from further heating by sunlight. The evaporation of these ices then virtually shuts down, and the comet becomes incapable of producing either a visible gas tail or a dust tail. The vast majority of the ice in the interior then remains hidden there indefinitely. The time required for the rest of the ice to evaporate and escape becomes much longer than the expected orbital lifetime of the comet, before it collides with a planet. In other words, the comet becomes indistinguishable from an asteroid, but its dusty exterior masks a massive heart of dirty ice. Water ice makes up about 50 percent of body's mass when water loss phases out. Further, a significant fraction of all short-period comets spiral in toward the Sun while they are expending the last of their surface ices, ending up in NEA-type orbits. Thus, the imposture is complete: the comet has taken on not only the surface appearance of a carbonaceous asteroid but also its orbit.

The theorists who studied the orbital evolution of comets, when provided with the information that comets can shut down their emission of gas and dust, came to a remarkable conclusion. The supply of transvestite asteroids from cometary evolution, combined with the supply from perturbations of real asteroids by Jupiter, adds up to a total supply rate that matches the calculated loss rate! Thus, about half of the NEA population consists of ice-rich inactive comet cores. This means that about a quarter of the mass in the NEA population is water. A wide range of other volatile materials is also present in all extinct comet cores and C asteroids.

Since the distinction between asteroids and comets in the NEA population is so difficult to draw, we shall henceforth refer to both groups collectively as near-Earth objects (NEOs). Because of their diverse origins and histories, NEOs are immensely varied in their compositions, ranging from natural stainless steel to cometary ice. Their potential as resources for humankind deserves careful study.

6

THE BEST THINGS
COME IN SMALL PACKAGES

"Steel" was a misnomer. The asteroid 8903 Steel was really a carbonaceous comet core, innocent of native metals. A fluffy, dusty black crust about an arm's length in thickness covered the surface with a velvet-black blanket. A few holes punched in the crust by recent meteoroid impacts leaked faint wisps of gas and dust where the probing rays of the Sun reached in to touch the ancient core of native ice.

The exploratory crew came with a few well-defined tasks. They were to land on Steel, assess its resources, and carry out several experiments related to future large-scale extraction of water. To do this, they needed to answer a number of basic questions: How much ice was present? How thick and how strong was the crust? How easy was it to extract water? How much of other useful materials, such as carbon, nitrogen, and sulfur, was present?

The exploration ship Anopheles, which actually looked more like a T2 bacteriophage virus than a mosquito, was specialized for its mission. Its long spindly legs and web-spinners were designed to hold fast to a small asteroid with almost no useful gravity. In the "nose" of the creature, aimed straight downward for drilling into the icy interior, was the "kucker" probe, designed to melt its way into the body and suck out its vital juices. A nuclear reactor rested right behind the kucker, in the "neck" of the mosquito, well situated either to heat ice and water for extraction or to heat liquid hydrogen for the nuclear thermal propulsion system to send them on their way home.

*Three cylindrical, heavily insulated tanks of liquid hydrogen sur-
rounded a central column that rose above the reactor. Atop the column
was a single large cylinder, the quarters where six crew members lived
crowded into three decks.*

*After twenty-four days on the asteroid, the plan went, the crew would
seal up a ton of samples and a ton of extracted water in an insulated,
refrigerated compartment beneath the lowest floor of the crew module
and depart on the return trip to Earth. About half the supply of liquid
hydrogen would be run through the reactor and heated into a furiously
hot stream of gaseous hydrogen for the rocket engine burn that would
lift them off the asteroid and inject them into an orbit that would take
them back to intercept Earth. When they arrived in the vicinity of Earth
a year later, a second nuclear rocket firing would use the remaining
propellant to kill their excess velocity and leave them in a highly elongated elliptical orbit about Earth. They would then be within easy
reach of an orbital transfer vehicle operating out of the new HEEO
transit base* Hermann Oberth.

*But the pressure relief valve on the hydrogen manifold had popped
off and stuck open shortly after landing, jetting a powerful stream of
liquid and gaseous hydrogen out at a sharp angle to the surface, tearing off two of the six fragile grappling legs and nearly ripping Anopheles from the asteroid. By the time the crew had diagnosed the problem
and dealt with it, over 80 percent of the hydrogen propellant supply had
been lost. There was no longer enough propellant to even get them back
to the vicinity of Earth, let alone safely captured into Earth orbit. The
crew lost its ticket home. To compound their hazard, two of the three hydrogen tanks were completely empty, effectively removing most of the
shielding that protected the crew from radiation emitted by the reactor.*

*The crew knew how to remove and replace the defective valve: they
had a redundant valve elsewhere in the hydrogen plumbing that could
be removed and installed with the tools at hand. The problem was that
the barn door had been open too long, and the horsepower was gone.
The powerful ten-megawatt nuclear reactor at the root of* Anopheles's
*kucker was as useless without its hydrogen as an automobile engine
without gasoline: all the thrust of the rocket came from the expansion of
hot gases heated by the reactor and directed by the rocket nozzle.*

As soon as it became clear that the geyser could not be shut off,

Pietro and Elisabeta Coradini, the chemical engineering/mining duo from ESA, Charley MacDougall, the Canadian trajectory and mission-planning honcho, and Sunny Lee, the nuclear reactor genius from Rocketdyne, gathered in the galley for a desperate brainstorming session. Tom and Andy Duncan, the mission specialist and pilot, the ones who really knew the vehicle hardware best, suited up and went out to look at the damage.

Twelve hours later Andy and Tom straggled in totally exhausted, the diagnosis and valve replacement already complete, and went to their cubby to crash. The brainstorming session went on for twenty hours without them. Some thirty options were suggested, then critiqued, then boiled down to just two possible courses of action. Option number one was a six-way suicide pact. Option two was to use the experimental kucker probe to suck a tankful of water out of the asteroid and put it in the liquid hydrogen tanks. Then they would try to use the water as the propellant for the return trip. The big questions were whether the necessary amount of water could be collected in time for the optimum departure date, and whether a nuclear reactor designed to heat hydrogen would survive exposure to hot water vapor. Extracting enough water was mathematically possible, but by no means a sure thing. Using the extracted water in the reactor was a complete unknown. Not only was the water corrosive, but dust and grit carried along by the water could jam or wear out valves, tear up the pumps, and sandblast the inside of the rocket thrust chamber.

The best solution they could think of was to pump the water through the reactor at the maximum possible rate. That kept the dirt moving and kept the reactor core much cooler than usual.

The kicker was that, in order to make room for a full load of water, they first had to dump the last of their liquid hydrogen overboard. Now there was an act of commitment.

So they did it. The kucker found so much ice that they filled all three tanks with mildly salty asteroidal water. The dust was fine enough so it was not a problem, but the salt threatened to be an even worse one. Running the reactor cool turned out to be a great idea. They departed from the asteroid on schedule. During the long coast back to Earth they went outside and looked at the engine from a safe distance. It looked OK. Not that it mattered; they would have to try to fire it again in any *case. Fortunately, the second engine burn worked fine. So there they were, back at Oberth with enough water aboard to fuel thirty-seven more trips to Steel.*

That is how the New Ice Age began. And that was the beginning of Tom Duncan's truly astronomical fortune.

—Mosquito

In *Rain of Iron and Ice*, I argued that the technology necessary to defend Earth against a wide range of possible comet and asteroid impact hazards is little beyond our present practice. Further, the capability to access these bodies for defensive purposes equips us with the ability to carry out scientific and engineering investigations of their surfaces and interiors. The step from scientific investigation and resource mapping to actual use of materials requires only that we know a single profitable scheme for extracting and using a single resource. Then access to all the other resources of near-Earth objects (NEOs) will be assured.

When we explored in chapter 4 schemes for using lunar materials, we were faced with a similar problem. What is the "foot in the door" that permits the development of lunar-materials processing? We found that two key resource uses and two basic technologies lay at the root of lunar industry. The early resource uses we identified were, first, shielding the occupants of a lunar base from cosmic and solar-flare radiation using screened but chemically unprocessed regolith, and second, making oxygen for life support and propellants. The corresponding key technologies were the ability to move and sort lunar regolith in a controlled manner and the ability to separate water into hydrogen and oxygen or carbon dioxide into carbon monoxide and oxygen, with the ability to reverse the process in the latter apparatus to generate electrical power from controlled fuel-cell recombination of the products. These abilities, modestly enhanced and combined, made large-scale oxygen extraction from the regolith possible via bulk regolith reduction using carbon monoxide or hydrogen and oxygen separation from the products. Slightly more advanced application of these basic processes would entail regolith or basalt crushing and screening, beneficiation of ilmenite, ilmenite reduction by reaction with carbon monoxide or hydrogen, and oxygen extraction as before.

It is important to realize that the driving considerations in developing both regolith-handling and oxygen-extraction processes on the Moon involve protecting the crew of a lunar base. But as we discuss NEO resources, we shall not assume the presence of permanent crew. Keeping the involvement of astronauts in asteroid mining to a minimum helps tremendously in keeping costs down. We shall instead consider the general utility of NEO materials in support of large-scale space activity, assuming that most early asteroid missions would be unmanned. This drives our attention strongly in the direction of propellant manufacture as the first step in NEO use. Providing life-support materials would take on added importance only after we have found out how to operate the propellant business profitably.

The highest-performance combination of rocket propellants is hydrogen and oxygen, which burn to make water. The optimum proportions of hydrogen and oxygen are thus two hydrogen atoms per oxygen atom, exactly the same as in water. The resource of choice for providing both fuel (hydrogen) and oxidizer (oxygen) is therefore water. At the temperatures and pressures prevalent on NEOs, the expected source of water can be either subsurface water ice or water-bearing minerals, such as clays or hydrated salts.

The equipment needed to extract water from an ice-regolith mixture (permafrost) includes a device for digging the permafrost, a pressure vessel heated by sunlight in which the permafrost is heated to distill off its water content, and a condenser to liquefy the water for storage. If we desire liquid hydrogen and liquid oxygen propellants, we may replace the condenser with a water-vapor electrolysis unit of the sort we discussed for use on the Moon, supplemented by refrigeration equipment to liquefy hydrogen and oxygen. The latter two pieces (of refrigeration and electrolysis equipment) account for the vast majority of the mass, complexity, power demand, and unreliability of the system.

Obviously, if we could use water itself as a propellant, we could enormously improve the plant's productivity and reliability. Engines that run on water are a traditional part of American culture. Many opportunistic fraud artists have collected "investment money" from naive "marks" for the avowed purpose of building automobile engines that defy the laws of chemistry and physics and run on water as their fuel. Of course, such investments do not have a good record of profitability for their investors. But is it in fact possible to run a rocket on water? The answer, astonishingly, is yes. Two different types of rocket engines can run on liquid water and eject steam, but they clearly must get the power to boil the water from somewhere.

The first kind of "steam rocket" is powered by a nuclear reactor. A stream of water is pumped into a hot reactor, where it boils and is heated to very high temperatures. The hot water vapor is vented through a rocket thrust chamber and imparts an impulse to the rocket. This *nuclear thermal rocket* can perform as well as a hydrogen-oxygen chemical rocket if the exhaust temperature is the same.

The second type of "steam rocket" uses a mirror to collect sunlight and focus it on the thrust chamber. Thus, it is solar power that provides the energy to heat the exhaust. This kind of propulsion system is called a *solar thermal rocket.*

A comparison of different kinds of rocket engines with each other requires some measure of their performance. The basic question is, if one kilogram of propellant from this rocket is fired so as to keep a constant thrust of one kilogram, how many seconds will the rocket operate before it runs out of fuel? A propellant that is very effective will produce a kilogram of thrust at a low propellant-consumption rate and will last a long time. A less efficient propellant combination will require a higher burn rate to produce the same thrust, and the kilogram of propellant will not last as long. This figure of merit for propellants is called the specific impulse, which (not surprisingly) is measured in seconds. A low-performance chemical rocket is one that has a specific impulse under 250 seconds, like the zinc-sulfur combination used in amateur rocket engines. A medium-performance propellant has 250 to 350 seconds specific impulse. This range includes the normal propellants used in liquid- and solid-fuel military rockets, including kerosene-liquid oxygen, hydrazine-nitrogen tetroxide, and rubber-ammonium perchlorate. A high-performance chemical rocket has over 350 seconds specific impulse. In practice, the only truly high-performance propellant is hydrogen-oxygen, which can achieve specific impulses of over 420 seconds when operated in space, where the exhaust can expand unimpeded and the exhaust velocity is highest.

It has been known since the time of Tsiolkovskii (chapter 2) that the specific impulse multiplied by the standard acceleration of gravity is just equal to the exhaust velocity of the rocket. Since the acceleration of gravity is 9.8 meters per second per second (this is not a misprint), the exhaust velocity of a hydrogen-oxygen rocket with a specific impulse of 420 seconds is 420×9.8, or 4,116 meters per second. The physical meaning of a high specific impulse is clear: what we really want is a rocket engine with the highest possible exhaust velocity. Physics tells us that the temperature of a gas is proportional to the product of the molecular weight times the square of its mean thermal velocity. Thus, the velocity increases with the square root of the gas temperature and decreases with the square root of the molecular weight. The best rocket exhaust is a very light, very hot gas. Carbon dioxide, a principal ingredient of the exhaust gases made during combustion of hydrocarbons, has a molecular weight of 44. Water has a molecular weight of 18, so it outperforms carbon dioxide at the same temperature by a factor equal to the square root of the ratio of 44 to 18, which is 1.562. At the temperature at which a hydrogen-oxygen rocket's exhaust (water vapor) has a specific impulse of 420 seconds, a carbon dioxide exhaust would have a specific impulse of only 269 seconds.

There are two practical limits enforced on chemical-combustion rockets. The first is that the temperature of the exhaust is limited by the materials out of which the thrust chamber is made: rockets become useless when they melt. The second is that the lightest molecule that can be made by combustion is water vapor. These factors force chemical rockets to deliver specific impulses well below five hundred seconds. There is another slightly more subtle limitation imposed by the fact that oxygen attacks most metals severely at high temperatures. Either very exotic and expensive metals or oxygen-free exhausts must be used to get to higher temperatures. Since it is combustion of a fuel with oxygen that usually gives it the highest specific impulse, this is a serious problem for chemical rockets. However, if we specify that we are using water as the "working fluid" that produces the exhaust stream, the maximum allowable temperature is determined by materials limitations, not by whether the energy for heating the exhaust came ultimately from chemical combustion, a nuclear reactor, or solar energy. The same ultimate specific impulse therefore applies to all three types of system. Thus, solar-thermal and nuclear-thermal steam rockets can perform as well as a hydrogen-oxygen chemical rocket *without the necessity of electrolyzing water and liquefying hydrogen and oxygen.* The penalty we pay for using a solar-thermal rocket is that we must carry a large mirror that acts as a solar collector. The penalty for using a reactor is that we must carry the weight of the reactor and its associated radiation shielding. But if we are using the reactor on unmanned missions only, the mass of shielding required may be very small. In both cases we also risk incurring the wrath of radical environmentalists who, uncomfortable with the fact that the entire planet Earth runs on nuclear power, assert that there is no such thing as a "safe" use of nuclear energy.

Does all this mean that we are forever stuck with a specific impulse limit of about 450 seconds? Well, what if we were to use the lightest possible gas, with no oxygen content—hydrogen itself. What could we achieve then? At the temperature at which water vapor (with a molecular weight of 18) achieves a specific impulse of 420 seconds, hydrogen (with a molecular weight of only 2), could achieve 1,260 seconds! The requirement that the exhaust be pure hydrogen (that is, contain no oxygen or water vapor) means that this exhaust cannot be heated by combustion. Where, then, would the energy to heat the exhaust come from? It could be heated by solar or nuclear power. Thus, enormous improvements in performance over chemical rockets could be achieved by solar or nuclear thermal rockets using hydrogen as the "working fluid" that makes up the rocket exhaust.

As a practical matter, hydrogen use has some negative effects. The engine-nozzle temperatures achieved to date permit specific impulses of 800 to 1,000 seconds. Reaching 1,200 seconds will require exotic heat-resistant alloys. Even so, 900 seconds is superb in a world in which 450 seconds once looked like a firm upper limit.

To illustrate the advantage of high specific impulse, let us compare the amount of fuel needed to accelerate a 100-ton payload to a speed of twenty kilometers per second. With a specific impulse of 1,260 seconds (hydrogen nuclear thermal or hydrogen solar thermal), the mass of propellant needed is 405 tons. If the specific impulse is limited by materials-engineering considerations to 840 seconds, the mass of propellant required is 1,036 tons. Finally, if the engine uses hydrogen-oxygen chemical propulsion or a nuclear or solar steam rocket, the specific impulse is 420 seconds and the mass of propellant needed is a staggering 12,700 tons. The latter mission is probably not even possible because the mass of tankage and structures needed to carry this vast mass of propellant exceeds the entire payload mass allowance. Thus, this mission is so demanding that we are forced to use liquid hydrogen as the propellant.

For missions with a much smaller velocity requirement, the differences in propellant mass requirements are much less profound. For example, consider a mission that requires a velocity change (called delta V) of only five kilometers per second. Then the propellant mass required at 1,260 seconds specific impulse is 50 tons, compared to 83.5 tons at 840 seconds and 237 tons at 420 seconds (the chemical-rocket and steam-rocket options). It therefore makes much less difference what specific impulse we use if the mission velocity requirements are modest.

In the real universe, the efficiency of the engine is a very important factor, but not the only one: the energy and equipment needed to process the propellants also strongly influence the overall performance of the system. We must also analyze these latter options in terms of the processing requirements for making the propellants for each of these low-velocity options. The steam-rocket mission requires that 237 tons of water be extracted. The hydrogen-oxygen chemical-propellant option (with the same specific impulse) requires that the same mass of water be processed, but also electrolyzed and liquefied, at enormous additional cost in energy and equipment. The intermediate specific impulse mission (840 seconds), reflecting current hydrogen thermal-rocket technology, requires the use of 83.5 tons of hydrogen. That in turn requires the extraction and electrolysis of 751 tons of water, plus the liquefaction of the 83.5 tons of hydrogen made by electrolysis. This option is therefore extremely wasteful of resources, compared to any of the three options with lower specific impulse. Finally, the highest-performance option, the advanced-materials hydrogen thermal rocket (1,260 seconds), requires 50 tons of hydrogen propellant. That in turn requires extraction and electrolysis of 450 tons of water and liquefaction of the resulting 50 tons of hydrogen. The conclusion is clear: for undemanding (low delta V) missions, lower-performance rockets are far better choices.

With this perspective on the propulsive uses of water and its ingredients, we can identify three different scenarios for extraction and processing. For operations in the region of space from Earth out to the asteroid belt, we need only extract water. The first option is to extract water from subsurface permafrost and use that water directly in a nuclear or solar steam rocket. The second option is to extract the water from ice-free minerals that contain water of hydration. Finally, for very demanding missions with high velocity requirements, we may choose to electrolyze water into hydrogen and oxygen and use the hydrogen in a thermal rocket, as just described. We have seen that the latter is very wasteful of resources, but, as in the missions with a twenty kilometers per second velocity requirement, as just illustrated, it may in some cases be the only way a very demanding mission can be done at all.

Water extraction from permafrost by distillation seems rather straightforward. Water extraction from ice-free rocks may seem a bit strange, but it has a strong foundation in our study of meteorites. Recall that the CI and CM chondrites contain from 5 percent to 20 percent water in the form of minerals containing water of hydration (H_2O) or hydroxyl (-OH) (see chapter 5). The CI chondrites contain abundant hydrated salts, such as the magnesium and calcium sulfates epsomite and gypsum, plus abundant clay minerals. Heating these materials to about a thousand degrees Celsius is sufficient to decompose these minerals and release the water contained in them. Other compounds are also released at the same time by decomposition of the abundant organic matter in them and by reaction of that organic matter with iron oxides to make water vapor and carbon dioxide. Nitrogen is also released by decomposition of the organic material, along with sulfur and sulfur oxides from the decomposition of sulfates and sulfides, and more carbon dioxide from the decomposition of carbonate minerals. The total amount of extractable volatile material can reach about 40 percent of the mass of CI material heated.

Initially, separating water from this mixture of gases by distillation will be the preferred route to a useful product. But later a demand for storable rocket propellants and life-support fluids could make the chemical processing of these gases very desirable. The familiar high-temperature electrolysis scheme for extraction of oxygen from carbon dioxide and water will give us a supply of carbon monoxide and hydrogen. From these materials, hydrocarbons such as methane can be made. With slightly different reaction conditions, the principal product can be methyl alcohol, which makes a very satisfactory middle-performance storable rocket fuel, especially when burned with liquid oxygen. The need for storable (noncryogenic, room temperature) oxidizers may encourage the production of hydrogen peroxide, nitrogen tetroxide, or nitric acid, which perform reasonably well as oxidizers and can be stored indefinitely without refrigeration.

The same mixture of gases provides nearly everything we need for life support: water, carbon dioxide, nitrogen, ammonia, sulfur, and so on provide the essential ingredients of organic matter. Phosphates, which are essential ingredients of nucleic acids and essential nutrients for all plants and animals, are also abundant in CI and CM chondrites.

The 20 percent to 40 percent of volatiles in CM and CI material contrasts very favorably to the 100 to 200 parts per million of solar-wind-implanted volatiles in the lunar regolith. If a carbonaceous NEO were as accessible as the lunar surface, its two-thousand-times higher concentration of volatiles would make it a two-thousand-times more attractive choice. But this is not the entire story. The solid residue left after baking the volatiles out of carbonaceous asteroid materials is of considerable interest in its own right. The reaction of organic carbon with magnetite consumes nearly twenty grams of magnetite per gram of carbon, which is sufficient to convert all the magnetite present to metallic iron. The solid residue left after autoreduction is about 30 percent ferrous metal alloy, with a composition very similar to that of iron meteorites. For comparison, the concentration of free (asteroidal!) metal in the lunar regolith is about 0.1 percent, some three hundred times lower. This means that asteroids generally have three hundred times as much free metal as an equal mass of lunar regolith—and the metal on the Moon is just debris from impacting asteroids that have exploded and diluted their metal down 300-fold! Why extract these materials on the Moon when there are plenty of nearby asteroids that are *easier to reach than the Moon, and three hundred times richer in metals?*

On average, the metal made by baking C asteroids (or already present in ordinary chondrite asteroids) will contain about 92 percent iron, 7 percent nickel, and nearly 1 percent cobalt. It also contains about twenty parts per million of platinum-group metals. These metals can be readily extracted in several ways for use as building materials, leaving behind a mixture of silicates of magnesium, calcium, aluminum, sodium, potassium, titanium, and other rarer elements. But water extracted from one batch of CI material can be used to leach out and recover water-soluble salts rich in sodium, potassium, halogens, and sulfates in the next batch before heating it. These leached materials are potential sources of a wide range of valuable industrial chemical agents such as acids, bases, halogens, and sulfur.

Combining these three steps, the fraction so far extracted for use would be 40 percent as volatiles plus 18 percent as metals and 3 percent of water-soluble leachate, leaving a dry, metal-free silicate residue making up less than 40 percent of the total mass. This material would at first probably be discarded or stockpiled for potential future use on an asteroid, since its value is too small to encourage exploitation. Ironically, the residual "slag" *discarded* after processing C-type asteroids is similar to average lunar surface material. Asteroid enthusiasts sometimes refer to the Moon as "the slag heap of the solar system."

When we discussed the use of lunar materials, we described how aerobrakes made of lunar refractories could enhance the effectiveness of lunar propellant manufacture by reducing the mass of propellant needed to slow down near Earth for rendezvous with the space station. Exactly the same kind of benefit would accrue from the manufacture of aerobrakes out of refractory asteroidal material to assist in returning asteroidal products to the space station or other locations in near-Earth space. It happens that the dry silicate residue that we suggested might simply be stockpiled during early asteroidal processing activity is rich in refractories. It is even possible that this residue could be used in its entirety to make heat shields. If so, the waste from this four-part processing scheme would be essentially zero.

Up to this point we have concentrated on the use of C asteroids or extinct comet cores because they are a conspicuously excellent source of propellants. But the compositional diversity of the NEO population makes all kinds of other resources available on other asteroids. The M asteroids are about 99 percent metal with the composition of a natural stainless steel. The E asteroids contain strange and readily extractable minerals of metals that have great metallurgical importance, such as titanium nitride, manganese sulfide, magnesium sulfide, nickel silicide, silicon oxynitride, and even elemental silicon. The latter three are potential sources of high-purity silicon for making thin-film amorphous silicon solar cells without the complexity and expense of extracting it from silicates. And if, for some unimaginable reason, we should decide that we want material similar to lunar basalts, the V and S asteroids can provide all we might need. Perhaps, after feasting on a few rich CI asteroids, our processing equipment will have to be put on a low-calorie diet of dry rocks for a while to unclog its arteries.

The resources in greatest demand for use in space are those that can be used to meet a space civilization's most massive material and energy needs. The cost of transportation from Earth is so large that any material needed in large masses in space, almost irrespective of its market price here on Earth, should—whenever possible—be made from resources found in space. Another corollary of this principle is that only materials with a high unit market value on Earth (thousands of dollars per kilogram) may be worth transporting back to Earth. Fortunately for us, the unit energy cost (kilowatt-hours per kilogram) of sending something back to Earth from any plausible mine site in near-Earth space will be less—and often enormously less—than the cost of launching from Earth. Earth, located at the bottom of a deep gravity well, is by its nature better suited to serve as an importer than as an exporter of raw materials.

As an example of the magnitude and economic value of space resources, we shall assay the *smallest known* M asteroid and account for its market value. That distinction belongs to the NEA known as 3554 Amun. Amun is only two kilometers in diameter, the size of a typical open-pit mine on Earth, with a mass of thirty billion (3×10^{10}) tons. Assuming a typical iron meteorite composition, the iron and nickel in Amun have a market value of about $8,000 billion. The cobalt content adds another $6,000 billion, and the platinum-group metals (platinum, osmium, iridium, palladium, and so on) add another $6,000 billion. Not counting the value of its major nonmetallic components (such as carbon, nitrogen, sulfur, and phosphorus) and its minor and trace nonmetallic components (such as germanium, gallium, arsenic, and antimony), the total market value of Amun in Earth's metals market is $20,000 billion. Since it is already located in space, it represents an asset that would cost about $10 million per ton to launch from Earth, a total of $300,000,000 billion. That is roughly equivalent to the gross global product of Earth for the next thirty thousand years. Therefore, this one small asteroid would provide us with the potential for a space program tens of thousands of times larger in scale than anything we could afford without the use of space resources—indeed, far greater than Earth's entire economy.

The alternative to economic exploitation is simply stated: we leave them alone. Of the two thousand kilometer-sized bodies in NEA orbits, about 4 percent (eighty bodies) are metallic or stony-irons. About a third of these (about twenty-seven bodies), if not intentionally deflected, will eventually collide with Earth, with devastating consequences. The impact energy depends on the impact velocity, but a body of this size in a typical NEA orbit would deliver at least a million megatons of explosive power. For comparison, the largest nuclear explosion ever detonated on Earth, a 1962 Soviet nuclear test on Novaya Zemlya, was just under sixty megatons. An all-out nuclear exchange exhausting the arsenals of the nuclear powers would deliver about twenty thousand megatons. Thus, one single impact of a kilometer-sized asteroid would deliver fifty times the maximum possible destructive power of a third world war.

Since there are about twenty-seven such bodies and since the mean orbital lifetime of NEOs is about thirty million years, we can expect roughly one iron-asteroid impact of this magnitude on Earth per million years. There will also be about three new recruitments of kilometer-sized metallic NEAs per million years into the NEA population.

It is reasonable to wonder when the next such impact will actually occur. At the moment, we do not know. But we now have the technology for efficient asteroid discovery. We could develop a nearly complete list of the large NEAs and determine their orbits within the next generation at very modest expense. We would then know what materials are available in space, which bodies threaten impacts, and when they will approach Earth. Equipped with this knowledge, we could both defend ourselves against the greatest natural threat to human civilization and take action to reap the riches of these extraordinary bodies.

The reader should not leap to the conclusion that, because I have here compared the option of using up an iron NEA with the option of letting it impact, the impact of irons of this size is a major part of the threat to Earth. Indeed, kilometer-sized bodies will penetrate the atmosphere and impact with the surface almost irrespective of their composition and strength. Nearly half of the impactors are probably similar to ordinary chondrites, and nearly half are carbonaceous bodies of asteroidal or cometary origin. Their explosive power is slightly less than that of an iron of the same size because they are less dense than iron. Of the two thousand NEOs larger than a kilometer in diameter, nearly seven hundred will eventually hit Earth. On the average, one body with an explosive power in excess of one thousand megatons (1 gigaton), will strike about every ten thousand years. The probability of an impact of this magnitude occurring between 1000 B.C. and 3000 A.D. is 0.4. This is certainly not a negligible threat.

Virtually all of these threatening bodies are attractive as resources. We have already discussed the carbonaceous and metallic bodies because they are rather clear-cut opposites that span the range of extremes between volatile-rich and metal-rich. But the ordinary chondrites are nearly as attractive sources of metals as the M asteroids. Ordinary chondrite asteroids contain an average of 15 percent to 20 percent metal in the form of small particles dispersed through silicate and sulfide rock. This stony material is certainly much easier to crush and handle than a massive chunk of stainless steel. Ordinary chondrites may prove to be an easier source of metals than an M asteroid. About 95 percent of the NEAs seem to be rich sources of either volatiles, ferrous metals, or specialty metals. The cost of dealing with the other 5 percent is trivial compared to the profits from dealing with the others.

Since some of the materials found in asteroids, most notably the platinum-group metals, are extremely valuable, it is sensible to ask whether they might profitably be returned to Earth. The former astronaut Brian O'Leary raised this question in the late 1970s, and Jeffrey Kargel of the U.S. Geological Survey has recently provided a tentative answer. He has analyzed the effects of importing large quantities of nonterrestrial precious and strategic materials on the market prices of commodities such as platinum and gold. Most important, he found no evident reason why such valuable materials might not be imported profitably if the importation rate is not too high.

Our analysis so far has emphasized the riches of lunar and NEO resources and the opportunities for exploitation of those resources. But no use will be made of even the richest nonterrestrial resources if the cost of reaching them exceeds their fair market value. Like all other endeavors so far attempted in space, the limiting factor on profitability of space resource use is transportation cost. Both the cost of delivery of equipment to the mine site and the cost of delivery of products to their place of use are important factors in the total transportation cost. It is time for us to examine the logistics of the acquisition and transportation of nonterrestrial materials to see whether there are any plausible scenarios for their use that do not incur burdensome logistic penalties.

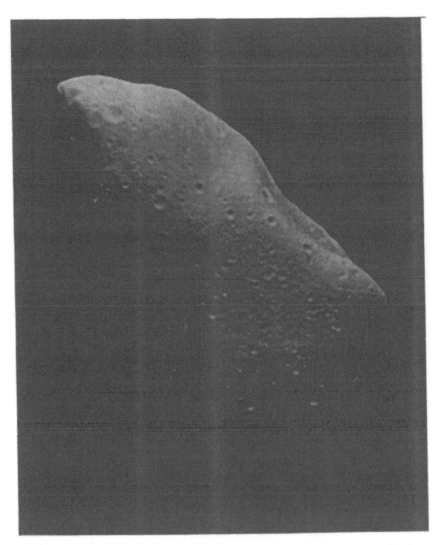

The Galileo spacecraft, en route to its encounter with Jupiter, flew close by the belt asteroid Ida and returned this picture of a surface heavily coated with regolith, shattered and spread about by countless small impacts. Photograph courtesy of NASA

This Viking Lander view of the surface of Mars shows how the neighborhood would appear to a visitor. Photograph courtesy of NASA

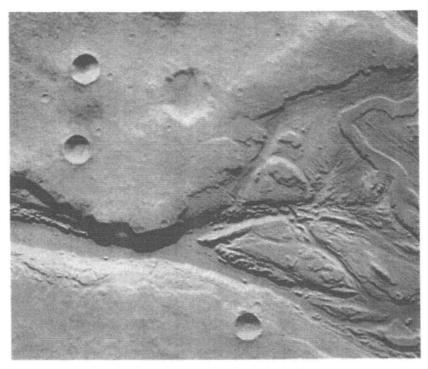

This Viking Orbiter view of a heavily eroded region of Mars shows enormous flood channels, some requiring water flow rates far larger than that of the Amazon River. Photograph courtesy of NASA

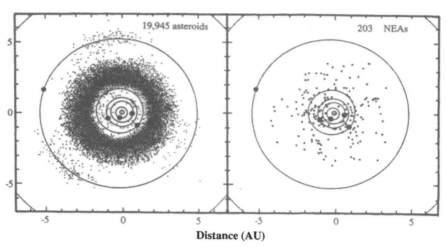

These diagrams show the instantaneous locations of all the known asteroids and of all the near-Earth asteroids. The famous gaps in the belt are not visible because certain orbital periods are forbidden, not certain instantaneous distances from the Sun. Courtesy of Mark Sykes, University of Arizona

Cooking volatiles out of an asteroid with solar power. Painting courtesy of David Egge

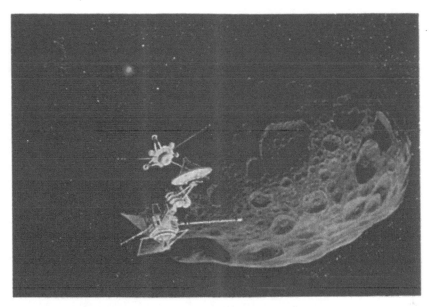

A pristine asteroid being surveyed for its chemical and physical properties and mineral content. Painting courtesy of David Egge

An emissary of mankind examines an asteroid. Painting courtesy of William K. Hartmann

Earth being visited by a large NEA (near-Earth asteroid). Painting courtesy of David Egge

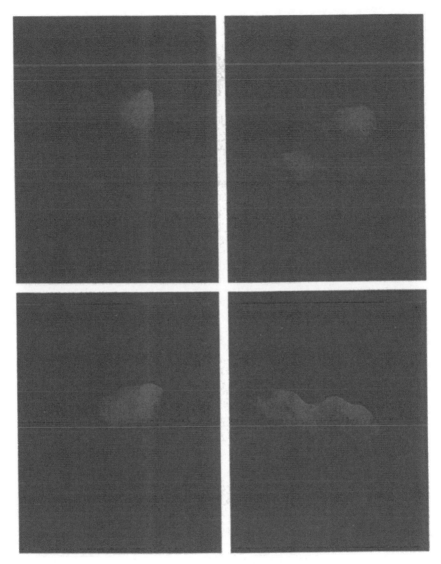

Four radar images of the near-Earth asteroid 4179 Toutatis, made by Drs. Scott Hudson of Washington State University and Steven Ostro of the Jet Propulsion Laboratory. These images, which also appeared in the October 6, 1995 issue of Science, (Vol. 270, p. 84), © 1995 by the AAAS, are reprinted here with the permission of Steve Ostro.

A vast cloud of habitats, sometimes called a Dyson Sphere, orbits far from the Sun to capture as much direct sunlight as possible. This view is looking Sunward from Pluto. Painting courtesy of David Egge

7

CATCHING A RIDE
IN SPACE

It took mankind thousands of years to realize that the Moon was the logical target for manned spaceflight. Once manned spaceflight had begun, it took less than two decades to realize that it was not.

—Metodius K. Pateyev
Vpered- na Mars!
Priroda Press
Novy Khabarovsk, 2077

READ THIS NOTICE BEFORE PROCEEDING FARTHER

Welcome aboard Cycler 3, *the third of six cycling spacecraft that operate on fixed orbits from Earth out to the inner edge of the asteroid belt.* Cycler 3, *like the other members of its family, was formerly a near-Earth asteroid whose orbit has been slightly modified to assure a metastable resonant relationship to Earth. Small amounts of propellant must still be expended from time to time to keep the orbit at resonance. All six cyclers have orbital periods of exactly two years and orbital inclinations of less than one degree relative to the ecliptic plane. Each cycler passes near Earth every two years, and a cycler passes Earth every four months. Although transport to the belt is the main function of the cyclers, they also make frequent close passes by Mars. Historically, at several times during the past three centuries the cyclers have been heavily involved in transporting Mars-bound traffic.*

Cycler 3 *is 800 meters long and averages 400 meters in diameter, except for the Earth-gravity wheel at the waist, which is 650 meters in diameter. Before reconstruction, the parent asteroid (2002 AK) was a carbonaceous asteroid of about half these dimensions and much less regular shape. The interior, originally mined to extract water for use as propellant to trim the orbit into resonance, has been further excavated and restructured to permit accommodations for 220,000 passengers at a high level of comfort and safety. Meteoroid, cosmic ray, and solar-flare proton protection provided by the outer walls are all well in excess of Earth-surface standards. Three centuries ago, each cycler accommodated as many as one million passengers per trip under trying and unhealthy conditions of extreme crowding. Most of these early travelers were emigrating prospectors, miners, religious or political refugees, space scientists and space technologists, and employees of essential service industries, plus their dependents. It was during this early era of mass emigration that the rules and procedures for operation of the cyclers as closed ecological systems were developed.*

The six Earth-Mars-Belt cyclers and the twelve intrabelt cyclers are operated by the Systemwide Cycler Authority, a nongovernmental, user-owned entity chartered to operate throughout incorporated space. Your own transit contract itemizes the details of the contractual relationship you have with the Authority. You are already contractually bound to adhere to all the clauses of your transit contract, including but not limited to those itemized below.

Experience has shown that, in order to preserve the structure and function of these cycling craft for future generations of users, it is essential that the following rules, excerpted from the full details given in the hard copy of your transit contract, be observed. Remember that there are no permanent crews on any cycler. Everything is up to you.

1. There must be no further extraction of water from the native (asteroidal) regolith on the surface of the cycler. The water content of the surface layer is an integral part of the shielding of the interior against light cosmic-ray primaries and energetic solar protons. Further water extraction jeopardizes the health and safety of the passengers.

2. All cycles of the biological elements hydrogen, carbon, and nitrogen MUST be closed. Certified manifests of all of these elements must be uploaded by all arriving vehicles. Departure manifests must also be

filed and reconciled before spacecraft-licensing documents will be re-
leased and before further flight plans can be approved. A related clause
of your contract specifies that any passenger who dies in transit must be
recycled on board. The other biologically essential elements—
including sulfur, phosphorus, transition metals, alkali metals, alkaline
earths, halogens, and nonmetals—are not presently controlled.

3. Users are contractually responsible for maintenance and repair of
the solar-cell arrays and farms. Highly automated facilities for repair
and recycling of failed units are provided.

4. All information necessary for running diagnostic, maintenance,
and repair functions is available as expert systems on TUTOR.

5. You are contractually forbidden to delete, edit, or extend system-
program or data files. All passengers are provided with secure private
directories for personal use. The health and safety of all passengers
could be severely compromised by any unauthorized change in this
software. The penalty for misuse of the computer system is immediate
recycling.

6. Orbital maintenance is under the control and supervision of the
SWCA. Security interlocks governing access to critical propulsion hard-
ware and navigation and control software can be overridden only at
SWCA authorization. One-time-only access codebooks are available by
Priority Red-1 emergency request only. The penalty for unauthorized
attempts to access said hardware or software is immediate recycling.

Advisory to Permanent Residents of Earth

Use of the Earth-gravity wheel is strongly advised but not contractually
mandated. Earth citizens have undisputed first priority for use of all 1-
g facilities. Visitors to Earth from other locations have second priority
inbound only. All others have no priority status but may access the 1-g
facilities on a space-available basis.

—Recycler

The resources of the Moon and asteroids are of no interest unless we
have access to them. In this case, access does not simply mean that
we can get there somehow; it means that we have technically feasible,

economically competitive transportation systems, with low propulsion energy requirements, and opportunities for repeated round-trips.

All outbound missions from Earth must pass through low Earth orbit (LEO). LEO is defined as the roughly circular orbit of an object about Earth at an altitude sufficient to keep it out of the atmosphere. In practical terms, a satellite in orbit below two hundred kilometers altitude experiences so much atmospheric drag that it cannot remain in orbit for many months. Space stations are so expensive that they cannot be placed in short-lived, unstable orbits. At about three hundred kilometers the drag is low enough that the station can remain in orbit for about two years. By firing small rocket engines from time to time, the space station can compensate for drag and lift its orbit to safer altitudes. About twice a year small supplies of fuel are delivered to the station for use in these orbit-maintenance maneuvers.

The Soviet Salyut and Mir space stations have been placed in orbit at about three hundred kilometers for even more pressing practical reasons: the Soyuz (A-2) launch vehicle, which carries the crew, life-support supplies, and orbit-maintenance propellants to their space stations, runs out of payload capacity when just above that altitude. American plans for an international space station have been based on the performance of the Space Shuttle. The shuttle system also pays a serious penalty for altitudes above three hundred kilometers but maintains a payload capacity adequate to reach six hundred kilometers. At that altitude, orbit maintenance is relatively easy. Further, the lower drag permits the use of large, flimsy, low-strength structures. Within these constraints, then, LEO means a circular orbit roughly five hundred kilometers above Earth's surface. A spacecraft in such an orbit has an orbital velocity of eight kilometers per second.

Missions heading outbound from a space station in LEO require a minimum velocity increment of 3.2 kilometers per second above circular orbit velocity to escape from Earth's gravity field. After a brief rocket-engine burn that adds a velocity increment (delta V) of 3.2 kilometers per second, the spacecraft will coast "upward," slowing down as it buys gravitational potential energy (altitude) at the expense of its kinetic energy (energy of motion). It will be traveling slowly three days later when it coasts past the Moon's orbit. If the Moon is nearby on its orbit, the spacecraft can match speeds with it at the cost of a further velocity change of several hundred meters per second. This would place the spacecraft in a high, nearly circular orbit about Earth that is the same as the orbit followed by the Moon. If executed close enough to the Moon, this maneuver can place the spacecraft on a collision course with the Moon. Over the next hours the spacecraft is in free fall toward the lunar surface, constantly accelerating in the lunar gravity. Then, with a single powerful engine burn, the spacecraft can decelerate to a soft landing on the lunar surface.

The delicate task of landing safely on a massive body requires that the spacecraft's altitude and velocity both reach zero simultaneously. These in turn require an accurate knowledge of the radius and gravity field of the Moon, or an intelligent onboard computer that uses a radar altimeter to monitor the vehicle's altitude and descent velocity during landing. The early failures in Soviet attempts to land Luna spacecraft on the Moon were caused by the fact that severe limitations on Soviet instrumentation and computer technology forced them to preprogram their landers. But the lunar gravity field was then so poorly mapped that the chances of achieving a successful preprogrammed landing were very small.

The delta V required by the spacecraft to carry out the orbit-matching and landing maneuvers, which may actually be done in a single engine burn, is just under three kilometers per second. The total delta V required to get from the space station to a soft landing on the lunar surface is about 6.0 to 6.2 kilometers per second, depending on the location of the landing site on the lunar surface.

The return trip from the lunar surface to the space station also consists of two parts. The first is an engine burn that lifts the spacecraft from the lunar surface, accelerates it to the lunar escape velocity and diminishes the orbital velocity about Earth inherited from the Moon. The spacecraft then enters a highly elliptical orbit about Earth, dives downward and grazes the top of the atmosphere at a speed 3.2 kilometers per second above local circular orbit velocity.

The second part is capture into an orbit that permits rendezvous with the space station. In chapter 4 we compared two scenarios for return to the space station, differing in how they dissipate their excess energy as they are captured into LEO. The first method, requiring a brief, powerful engine burn as the spacecraft streaks by the space station, is the mirror image of the departure burn by which it originally left LEO. The other method is to dissipate the extra 3.2 kilometers per second by passing through the upper atmosphere. A refractory aerobrake heat shield protects the spacecraft from the frictional heat. If the return to the space station from the Moon is done entirely by rocket propulsion, the total propulsive delta V requirement is exactly the same as for the outbound trip, about 6.1 kilometers per second. But if the capture into LEO is done by aerobraking, the propulsive energy requirement is reduced by 3.2 kilometers per second. After aerobraking, the spacecraft is in a slightly elliptical orbit with a perigee of about 100 kilometers altitude and an apogee near the space-station altitude. It will then require a minor rocket-engine burn of about 100 meters per second to lift the perigee out of the atmosphere and match orbits with the space station. The total propulsive delta V for aerobraking return from the Moon to LEO is then 6.1 − 3.2 + 0.1 = 3.0 kilometers per second.

Scientists have conducted several studies of the use of lunar propellants (principally lunar liquid oxygen) to support the transportation system that links the space station with a mature lunar base. In general, there must be a manned and relatively complex lunar base that supports not only a scientific crew but also a mining and processing team dedicated to the manufacture of liquid oxygen from lunar minerals. The usual system design has a dedicated lunar-landing and -takeoff vehicle that shuttles back and forth between the lunar surface and low lunar orbit (LLO). The absence of an atmosphere on the Moon permits orbits almost as low as the mountaintops. However, the lumpiness of the lunar gravity field constantly perturbs the orbit of any spacecraft in LLO, and some safety allowance must be left to preclude an unfortunate collision with a mountaintop at an orbital velocity of 1.7 kilometers per second. Because the lander has legs, landing footpads, and a suspension system that are useless for the remainder of the mission, they are either stored at a depot in LLO, or the cargo (crew members being rotated back to Earth, scientific samples, an aerobrake heat shield, and tanks of liquid oxygen) is transferred to a dedicated transfer vehicle that cycles between LLO and LEO, taking three days of flight time in each direction. Part of the lunar oxygen is burned with liquid hydrogen (brought up from Earth) to transfer into an orbit that grazes the upper atmosphere, and the lunar-manufactured heat shield is used for aerobraking. A small amount of additional propellant is expended to rendezvous with the space station, and the rest of the lunar liquid oxygen is delivered to a refueling facility at the space station, where it is available for use in sending the next mission out to the lunar base.

The first load of lunar oxygen returned to LEO is small recompense for the great mass of equipment that must be launched from Earth to extract and return the oxygen. But the processing and transportation equipment is now in space and can be used extensively. Each succeeding trip returns tons more lunar oxygen, but each trip requires a small mass of hydrogen shipped up from Earth via the space station.

After many cycles, the cumulative mass of material delivered to LEO is larger than the mass that was originally launched from Earth. For a fully mature system, which uses very reliable long-life components and has long since paid for the installation of the necessary hardware at the lunar base, in LLO and the space station, the ultimate leverage that can be achieved is a return of 2.4 tons delivered to the space station for each ton originally launched from Earth. That number is based on elaborate computer models of the complete processing and transportation system done by workers at NASA's Johnson Space Center and elsewhere. We say that the *ultimate mass payback* of this scheme is 2.4:1. Whether such a small mass payback is *economically* profitable is far from obvious. The cost of these complex operations is considerable but poorly known. Some analyses claim a net economic benefit; others do not.

If hydrogen were available on the Moon, then the mass needed to be brought from Earth would be significantly reduced, and the mass payback would be greatly improved. But the only plausible near-term source of the needed amounts of hydrogen on the Moon is polar ice. Logistic considerations make the Moon a very attractive base of operations if ice is abundant in the lunar polar regions. But if the principal base of lunar operations is in the equatorial region, the attractiveness of polar ice is greatly diminished. It is not easy to get the water from the polar regions to the equatorial base, where it is needed. The minimum distance from pole to equator is 2,740 kilometers, but on the tortuous, shattered lunar surface it is questionable whether a navigable surface route shorter than 4,000 kilometers could be found. It would require about a thousand hours of driving time to make the trip on an unimproved track, which is three lunar months' worth of daylight. The amount of fuel needed to drive this route is probably beyond the carrying capacity of the vehicle. It would be much easier, and probably much safer, to transport the payload ballistically, using a rocket that is launched from the pole with sufficient fuel for a round-trip to the equator and back, plus payload to be left at the equatorial base.

The use of polar ice to support such a ballistic-transportation scheme requires a substantial consumption of propellant along the way. Suppose that we have a 100-ton vehicle, consisting of 88 tons of tankage and structures plus a 12-ton crew compartment. At the pole, we fill the tanks with 400 tons of water and 700 tons of hydrogen and oxygen. The vehicle blasts off from the pole, coasts a quarter of the way around the Moon on a ballistic trajectory, and fires its engines again to soft-land at the equatorial base. This trip, taking less than a half hour, consumes 600 tons of propellant. The landed vehicle then offloads its 400 tons of water, leaving only 100 tons of hydrogen and oxygen in its tanks. The vehicle finally blasts off on its return trajectory and lands at the polar base at a gross weight of 100 tons, with nearly empty tanks. The resources consumed amount to 1,100 tons of water, plus the energy needed at the polar mine site to electrolyze 700 tons of water and liquefy the resulting 700 tons of hydrogen and oxygen. Of the 1,100 tons of water mined, 64 percent is expended to transport the remaining 36 percent to the site of the equatorial base.

My conclusion is that export of propellants from the Moon for use in supporting space transportation is economically very marginal unless polar ice is found in quantity. In the absence of ice, export of metals and propellants from the Moon is of marginal attractiveness because of the Moon's substantial gravity. For a commodity like steel or water, the cost of providing this material from Earth is dominated by its launch cost. On the Space Shuttle, that cost is close to $10,000 per kilogram. On a commercial launcher a few years hence, that cost should be closer to $1,600 per kilogram, and there have been several suggestions for future launch systems that could conceivably deliver payload to orbit for as little as $300 per kilogram. The challenge, then, is to deliver lunar materials to the space station with a total operations cost under $300 per kilogram. With a mass payback ratio of 2.4:1, each investment of $300 in launching 1 kilogram of equipment eventually results in placing 2.4 kilogram into the Space Station orbit. The cost of lifting that mass from Earth would be $720. Thus, the cost of processing and transportation for lunar resource use must be less than $420 per 2.4 kilogram ($175/kilogram). Failing that, export of lunar materials will not make economic sense.

The near-Earth objects present a very different set of problems. The richness of NEO resources cannot rationally be doubted. Typical desirable resources are hundreds to thousands of times more abundant in NEOs than on the Moon. But the round-trip times for missions to NEOs are typically two to five years—not one week, as with the Moon. The logistics for NEO missions are also very different, in part because they pursue independent orbits about the Sun, and in part because their very low surface gravity makes landing and takeoff very easy. Further, the potential availability of asteroidal propellants complicates the assessment of schemes for using lunar resources by raising the prospect of cheap water (and hence cheap hydrogen and oxygen) at the space station. Thus, asteroidal resources may effectively make the Moon more accessible from Earth.

The delta V requirements for a number of missions to dozens of NEOs have been calculated in the course of planning future one-way and round-trip (sample return) scientific missions. The results are most interesting: 20 percent to 25 percent of the NEOs are astonishingly accessible from Earth. A given rocket booster could actually land more mass on these bodies than it could land on the Moon! Of course, that is not the only factor that affects the ability of a mission to return mass to the space station or other orbit about Earth. Another very important factor is the *return* delta V requirement, since it is on that leg of the trip that the mass of the returned material must be moved. For return of material from the Moon, the delta V requirement is dominated by the need to break free of the Moon's gravity field. Asteroids, being very much less massive than the Moon, are vastly easier to escape from.

It is interesting to compare the results of sending a single round-trip mission from the space station to the Moon with a similar mission to an NEO. Both missions depart from the space station using hydrogen-oxygen chemical propulsion. In the NEO case, we shall assume that water is available on the target body for use in the return-trip propulsion system, and that aerobraking is used for capture into LEO. For the Moon, we assume the availability of lunar liquid oxygen. For missions to the Moon, the outbound delta V is 6.1 kilometers per second and the inbound delta V is 3.0. For a "good" NEO, we shall use an outbound delta V of 5.0 kilometers per second (the minimum possible is 3.3, and the smallest known is 4.3 kilometers per second) and an inbound delta V of 1.0 kilometers per second (the minimum possible is zero, and the smallest known is 0.06 kilometers per second).

Now, to compare the logistics of return of asteroidal materials, let us review the two principal lunar cases (without and with lunar polar ice). The first lunar round-trip mission incurs the cost of launching the spacecraft, plus all the fuel needed to land it on the Moon, plus the hydrogen needed for the return from the Moon to the space station (and the fuel to land that hydrogen on the Moon!). Not surprisingly, even after the first mission gets back to the space station with its load of lunar oxygen, no advantage has yet been realized. But thenceforth the spacecraft is already in space. The additional launches from Earth are simply those required to provide the modest mass of fuel for the lunar landing and blastoff (all the liquid hydrogen in this scenario must be lifted from Earth). Each successive round-trip to the Moon then returns more mass to LEO than must be lifted from Earth to keep the operation going. The mass payback soon rises above 1:1 and after several missions is showing a significant mass advantage. Assuming that the spacecraft lasts a very long time, the mass payback ratio eventually grows to 2:1 or 2.4:1. Whether this scenario is economically profitable is, as we have seen, certainly debatable.

If we modify this scenario to take into account the availability of lunar polar ice at the lunar base, we no longer have to land terrestrial hydrogen on the Moon and pay the transportation penalty for bringing it. All the hydrogen and oxygen consumed during takeoff from the Moon is made out of lunar water. Furthermore, the continuing need to launch propellants from Earth vanishes when there is a supply of water arriving from the Moon. With these advantages, the mission actually breaks even on the first flight. The mass payback continues to improve with each additional mission. After ten round-trips the mass payback ratio is around 8:1. It seems very likely that this scenario would be economically self-supporting if the lunar base is located near an ice deposit in the lunar polar regions.

A similar scenario for a round-trip mission to our typical good NEO would yield a mass payback of about 3:1 for the first trip and about 15:1 after five round-trips. Interestingly, if these missions are based in a highly eccentric orbit about Earth, rather than in LEO, the mass payback at least doubles. Several cases have been identified in which, after five round-trips from highly eccentric Earth orbit (HEEO) to the surface of a water-bearing NEO, we have achieved a very attractive mass payback of 100:1. This is numerically far superior to the lunar cases, even the one in which lunar polar ice is present and abundant, but the time required to realize this advantage is years, not months. Perhaps lunar ice should be kept on the Moon for use by its residents.

The payback improves if the asteroidal spacecraft is capable of numerous round-trips before expiring. The number of trips that can be expected for a spacecraft that transports water back from an asteroid is difficult to estimate. If we assume a lifetime similar to that of commercial communications satellites (which are much more complex), we can expect about twenty years of operational lifetime before a crippling failure. Considering that round-trip missions to NEOs take three to six years, we can expect about three to seven round-trip missions within this time. There is, however, an important difference between the two types of missions: communications satellites in geosynchronous orbit (GEO) are utterly inaccessible from the day of insertion into orbit until they expire, whereas the asteroid vehicle returns to a manned space base every few years for preventive maintenance, diagnostic testing, and repair. When an asteroid return vehicle approaches retirement age, replacement of a very small portion of its total mass— such as its attitude-control thrusters, sensor optics, pumps, and mechanical moving parts—with Earth-fabricated components might restore the spacecraft to full function. Modularity of design and ease of service and component replacement are essential. In that case, spacecraft may survive for many decades, and ultimate mass payback ratios of 1000:1 would be achievable.

Whatever the spacecraft lifetime, it is very significant that each asteroid round-trip provides twenty to fifty times as much propellant at the space station as would be needed for the next outbound trip to the surface of a nearby asteroid. Thus, each trip makes considerable masses of propellant available for other uses in near-Earth space.

Asteroid retrieval missions can be made more profitable by several other factors besides using HEEO as a base. The use of a heat shield made from asteroidal materials is highly attractive because it reduces propulsion requirements. If the propulsion system for return from the surface of the asteroid is a solar or nuclear steam rocket, then electrolysis of water and liquefaction of hydrogen and oxygen at the asteroid need not be attempted, with consequent major savings in equipment mass and processing power. Further, only water need be transported on the return vehicle. There are no cryogens aboard (and hence no need for refrigeration equipment), and propulsive accelerations are very mild. It then becomes possible to carry the water as a large "ice cube in a plastic bag," with a "tank" that weighs in at less than 1 percent of the mass of payload, not the 8 percent more typical of liquids that require support, confinement, and insulation.

But these numbers have a second implication. A spacecraft returning from an asteroid is not limited to carrying water. It may devote a portion of its large payload capacity to carrying unprocessed permafrost, carbonaceous chondrite solids, metallic iron from autoreduction of the carbonaceous chondrite material, or any other commodity available on the asteroid. Alternatively, we may use one mission to a carbonaceous or cometary NEO to provide enough propellant for a number of missions to metal-bearing NEOs to provide a massive supply of metals for construction use in any desired orbit in near-Earth space. Construction projects in Earth orbit that require great masses of metals may become economically feasible only through the use of asteroidal metals and propellants.

With future launch costs from Earth in the vicinity of $1000 per kilogram, this scheme could make asteroidal materials such as water or metals available in orbit around Earth for as little as $1 to $10 per kilogram. Ten tons of raw material, sufficient to make a space habitat for one person, would then cost about $10,000 to $100,000, compared to today's $100,000,000. Combining a plausible 10-fold reduction of launch costs from Earth with 1000:1 mass payback from asteroids makes a *10,000-fold improvement* in the affordability of space.

In a broader view, the use of propellants made on the Moon or asteroids greatly extends the range of attractive missions to the Moon, Mars, the asteroid belt, and elsewhere by dramatically lowering Earth launch costs. The main function of the Moon, with its low resource quality and high escape velocity, is to defray the cost of operations on, to, and from the Moon itself. The asteroids afford much more powerful leverage—but only if we plan on time scales of decades. The NEOs then serve as refueling stations and stepping stones that serve the needs of all operations in space, from the space station outward.

8

POWER FROM THE SUN,
THE MOON, AND ASTEROIDS

His heart pounding, Sessue Nakazawa gazed out the tiny window of his room at the International Lunar Base. He breathed deeply, seeking composure for the task at hand. When his breathing steadied, he turned on the tiny recorder in his hand and began to speak.

"Father! At last I am here! Since I was a child, I have dreamed of the time when I could stand on the Moon, when I could help in the great worldwide effort to find clean, abundant new energy sources for Earth. I know that it has been hard for you to see me depart so far from your plans for . . . " *Nakazawa hesitated, stopped, and rewound the tape to erase the last sentence.*

"Your encouragement to find a worthy career goal that can bring honor and wealth to our family has helped carry me through my years of study. It was hard for me to be away from my family and your guidance for so long." *He smiled to himself and shook his head slightly before continuing. My years of study at the Space Engineering Research Center at the University of Arizona and the Center for Space Power at the University of Houston have prepared me well. But now I am on my own for the first time."*

He paused the recorder to collect his thoughts. Directly in front of his window, and extending outward from it for over a kilometer, the lunar surface had been plowed smooth. At the near end of the leveled strip, close enough to the base complex to be connected by an unheated but vacuum-tight pedestrian tunnel, was the LUNOX Corporation plant, which extracted oxygen from lunar minerals to provide air for the base and oxidizer for rockets departing from the Moon. A newer-looking module attached to the LUNOX facility contained the Ludwig Mond carbonyl plant for extracting and purifying iron, nickel, and cobalt from the LUNOX byproducts and from impure metal grains extracted from the lunar soil by magnetic rakes. "As I look out my window, I see the beginnings of vigorous, new growth on the Moon. The Moon is no longer dead."

The fate of iron from the Mond plant was obvious to the eye: a spidery framework of thin iron girders marched halfway down the leveled strip to where a construction robot could be seen adding a shiny new length of girder. Another nearby structure, still quite small but slated for considerable expansion, stood tall and narrow at the right-hand edge of Nakazawa's field of view. That was the Criswell Corporation's silicon plant, where high-purity silicon tetrafluoride gas made from lunar dirt was processed into thin amorphous-silicon solar cells. Soon the steerable array of girders would be complete, and the delicate task of attaching vast expanses of solar cells would begin.

Though Nakazawa was thrilled by the idea of the lunar power station, he had already seen some similar hardware being built in geosynchronous orbit on his way out from Earth. As he had rested in the cramped passenger quarters of the HEEO base Goddard, waiting to connect with his outbound flight from Space Station Alpha to the Moon, he had seen a similar but even vaster array of solar cells under assembly.

"The Moon is coming alive under the hands of men and women from Earth. All Earth will benefit." He paused again to consider his wording. With a wry grin he continued, "The prospects for profit are enormous, of course, but there are other benefits as well."

The real reason for his trip to the Moon was almost entirely out of sight from his tiny window. The helium-3 depot was just visible, but the vast expanse of strip-mined lunar dirt, out of sight to his left, was clear in his mind's eye. The landing pad from which the helium-3 tankers departed was empty at the moment: somewhere between Earth and Moon two small tankers, one outbound to Earth with its precious cargo and one inbound with its helium tanks empty, tied the economies of Earth and Moon together by a frail lifeline. Nakazawa reflected that over

1 percent of Earth's energy needs were already being provided from the Moon. His job, over the next ten years, was to make that 25 percent. He stood straighter and spoke out in a louder and more formal voice. "Father, I hope to make you proud of me." As he turned from the window, his voice caught in his throat and he could not speak for several minutes.

—Lunar Power!

Which is worth more—a pound of gold or a pound of light? The answer is light, hands down.

Of all the commodities that we might hope to transport back to Earth from space, the one with the highest value per pound is unquestionably energy. A large fraction of the industry of Earth is devoted simply to mining, refining, and transporting various sources of energy, such as crude oil, natural gas, coal, and uranium. Certainly, the cost of energy is not high enough to justify transporting hydrocarbons from another body in the solar system to Earth: the energy used to transport such low-value material would greatly surpass the energy content of the material transported. The only kind of energy source that might be worth returning to Earth is one with a sufficiently high value (dollars' worth of energy content per kilogram) to justify the cost of shipping it to Earth. Indeed, in the purest form, energy need not weigh anything at all! If we could transmit the energy itself to Earth, we could eliminate the need to return massive tonnages of fuel at vast expense. Perhaps we could find some very concentrated source of energy in space, or even find some means of sending energy from space to Earth.

In space, the first form of energy that comes to mind is solar power. The Sun is widely used to generate electrical power for satellites in low Earth orbit (LEO), even though these satellites are in Earth's shadow about 40 percent of the time. Following a suggestion made in 1945 by Arthur C. Clarke, commercial communications satellites are usually placed in high circular orbits above the equator, at an altitude of about forty thousand kilometers, at which their orbital period matches the length of Earth's day. Since these satellites always stay above a fixed point on Earth's equator, the path they follow is called geosynchronous orbit (GEO). A satellite in GEO experiences perpetual sunlight except for brief eclipses lasting a few minutes to seventy-five minutes and occuring once a day for about eighty days a year. Indeed, a satellite in GEO is in unobstructed sunlight more than 99 percent of the time, averaged over the year. A working system of these power satellites would consist of many such units spread about Earth's equator in a "constellation" of satellites.

Beginning a quarter of a century ago, Dr. Peter Glaser, of ADL Inc. (previously known as Arthur D. Little, Inc.) in Cambridge, Massachusetts, advocated the use of GEO as a base for huge arrays of solar panels to convert sunlight into electricity. These *photovoltaic arrays*, called solar power satellites (SPSs), may have several square kilometers of solar cells exposed to direct sunlight in space, which has an intensity of about 1,300 watts per square meter. If these solar panels have an overall efficiency of 20 percent (some laboratory cells have produced as high as 30 percent), then each square kilometer of solar-cell array generates 260 megawatts of electrical power.

Largely as a result of publicity generated by Glaser, NASA undertook a study of the economics of SPSs in 1977. The ground rules were simple: A new advanced launch vehicle with very low launch costs (a few hundred dollars per kilogram of payload delivered to LEO) was assumed to be available. The entire SPS system would be launched in prefabricated units by large numbers of these launch vehicles. A construction crew would travel to GEO to assemble these units into a working SPS.

Each completed SPS would be a huge flat array of solar panels kept perpetually aimed face-on at the Sun. At GEO, atmospheric drag is nonexistent, so fuel use for orbital station keeping and panel-orientation adjusting is very low. The electrical power generated by the array runs a powerful microwave transmitter that aims a narrow, precisely aimed beam, which illuminates a flat array of antennas covering several square kilometers of ground on Earth. The antenna array receives the microwave power beam and converts it with very high efficiency into electrical current, which is then fed into the power grid.

Economic analysis of this scheme, however, showed it to be very marginal. The lowest conceivable launch costs and the highest plausible solar-cell efficiencies and lowest solar-cell weights foreseeable in the late 1970s combined to suggest that an SPS could not deliver power at the present price of electricity. True, we may elect to pay two or three times as much for electricity if our present sources of power are used up—but until that time, SPS systems are not economically competitive.

From the vantage point of the mid-1990s, several factors in this analysis look very different. Research on solar-cell design has recently shown us how to make extremely lightweight solar cells. These new lightweight cells use a very thin layer of amorphous (non-crystalline, or glassy) silicon instead of the thick slices of crystalline silicon used in conventional solar cells, saving great amounts of weight that would otherwise have to be lifted from Earth to GEO at a cost over five hundred dollars per kilogram. Also, research on robotics and automation has suggested several ways to minimize, or even eliminate, the participation of astronauts in assembling and installing SPS constellations, saving even more on launch costs, life support, radiation shielding, and so on.

Another insight has arisen out of studies of the logistics of deep-space missions that must return to orbit about Earth. These studies have consistently shown that the ideal orbit for the base of operations is a highly eccentric Earth orbit (HEEO), reaching in as close as 1,000 or 2,000 kilometers altitude at perigee, and as far as 100,000 to 400,000 kilometers at apogee. Launch costs from Earth are about twice as high to HEEO as they are to LEO—but both the cost of Earth-escape missions from the station to the Moon, NEOs, or Mars and the cost of return to the station are dramatically reduced. In fact, HEEO is not any harder to reach than GEO, but it is *much* easier to get back from. Therefore, a HEEO base is a very attractive staging area for the retrieval of asteroidal metals and propellants, the fueling and launching of planetary exploration missions, and the construction of large permanent space structures. The need to keep the transmitting antenna on the SPS centered on the receiving antenna on Earth makes the HEEO scenario more complicated than the GEO one, but these problems mostly involve electronic beam-steering and computer-control technology, both of which have advanced greatly since the 1970s.

As the 1977 study showed, building a constellation of SPSs in

HEEO makes no economic sense if we must pay the cost of lifting all the construction materials of the SPS from Earth's surface. But if we can instead make most of the mass of the SPS from nonterrestrial resources, the logistical advantage of HEEO becomes a powerful positive factor. But does it make sense to fabricate SPS components in space out of nonterrestrial materials? In my view, returning lunar or asteroidal materials to a factory at HEEO gives us enormous leverage. Each ton launched from Earth can provide us with 100 tons of asteroidal metals at HEEO for use in SPS construction. The challenge of *how* to use that powerful leverage for the benefit of humankind is simply an intelligence test. If we pass the test, we will be able to provide unlimited amounts of energy to Earth at costs below current energy costs, without further fossil-fuel consumption or nuclear-power-plant construction. If we flunk the test, we get to freeze in the dark.

We could, for example, make the low-tech parts of the SPS satellites (wires and cables, propellants, beams and fixtures, even solar cells) in space if transportation costs from the Moon or nearby asteroids are low enough. A lunar-based scheme would entail using lunar liquid oxygen to lift metals and silicon off the lunar surface and to inject them into HEEO. The ferrous metals may be fabricated into finished SPS components on the Moon or at an automated facility in HEEO. The propulsion requirements for this mission are almost identical to those for the aerobraking return to LEO discussed earlier, except that no heat shield is required (because there is no longer any need to slow the vehicle at perigee in order to drop into LEO). HEEO, which reaches out almost to the Moon, is simply much more accessible for round-trips from the Moon than LEO is. If the destination is HEEO instead of LEO, export of material from the Moon makes much more economic sense.

HEEO is also much more accessible than GEO for all return missions from asteroids, requiring much less propulsion energy, whether or not aerocapture is employed. Thus, asteroidal materials also are most advantageously returned to HEEO.

Note that, from LEO, HEEO is accessible at a delta V cost of only 3 kilometers per second, compared to 6 kilometers per second to reach the lunar base. This means that each launch from Earth can deliver more than twice as much payload to HEEO as to the Moon. Further, starting from a space station in LEO, HEEO is actually more accessible than GEO (3 kilometers per second versus 4 kilometers per second) because there is no need for an orbit-circularization engine burn in HEEO, as there is entering GEO.

We should recall here from our earlier discussion (chapter 4) that a lunar transportation system that uses lunar oxygen and terrestrial hydrogen is logistically more complex and has less capability than one based solely on lunar polar ice. The best prospect for profitably exporting massive amounts of metal from the Moon would be to use lunar polar ice to make the propellants and to use a base of operations in HEEO.

There is, however, no question that the use of propellants and metals from NEOs can provide a very large leverage factor. If SPSs are constructed in HEEO with the maximum possible use of nonterrestrial resources, then it is useful to design them in such a way that the mass of terrestrial material is minimized. This minimizes the mass of material that must be lifted from LEO to HEEO. Since HEEO is already more accessible than GEO, this is a less urgent matter than if we were compelled to operate in the less accessible GEO environment. Further, transporting the Earth-built components of an SPS to HEEO can be accomplished using asteroidal water returned to LEO rather than terrestrial propellants for the LEO-HEEO leg.

A second Moon-based solar-power scenario, suggested by David Criswell of the University of Houston, extracts several of the strong points of the lunar-materials scenario while avoiding some of its deepest pitfalls. In essence, Criswell proposes to build solar voltaic collectors out of lunar materials—and install them on the Moon. The entire economically problematical process of exporting vast masses of material from the Moon is neatly avoided. The power generated on the Moon is then transmitted by microwave beam back to Earth. In one imaginative sweep, Criswell has done away with the (possibly futile) requirement for accessible lunar polar-ice deposits and removed the need to launch tens of thousands of tons of metal off the Moon. The metal, and even the silicon to make the solar cells, is used close to where it is mined and processed. Solar-powered lunar trucks replace fuel-guzzling rockets. (It is amusing that the lunar-power-station scenario would work just about as well without lunar polar ice as it would if there were abundant ice.) These enormous advantages are only partially offset by two negative factors. First, any solar array on the lunar surface shuts down during the 50 percent of the time that it is local night, and a solar array is vulnerable to thermal stress damage during the cooldown and warmup. For two weeks at a time, each lunar power station is out of service. Near the time of new moon there is no point on the Moon that can both see sunlight and transmit to Earth. Second, transmission of beamed microwave power to receiving antennas on Earth is somewhat more difficult than from HEEO, which is in turn more demanding than from GEO. Neither of these seems to be a show-stopper, but both factors must be taken into account in a detailed economic analysis.

Construction of a string of power stations about the lunar equator, linked by a Moon-girdling equatorial power cable, would do away with the new moon dead-time problem, but of course the duty cycle of the system would still be 50 percent. To my mind, the most serious challenge is to minimize the cost of establishing the smallest possible profit-making power system.

I have heard critics of this idea say that it is dependent on the existence of a highly capable lunar base and is therefore unlikely. My response has been that this scheme finally provides a credible rationale for the existence of a highly capable lunar base, because it is the first scheme that seems capable of paying its own way.

The Moon also figures centrally in another scheme to meet Earth's energy needs in the coming millennium. As Gerry Kulcinski and his co-workers at the University of Wisconsin have noted, the lunar surface contains a trace of the light isotope helium-3, which is implanted in the surface by the solar wind. This isotope, which is extremely rare on Earth, is of interest because helium-3 is a potential fuel for use in fusion reactors.

Fusion of light isotopes is the energy source that powers the Sun and the stars. In general, stars fuse hydrogen into helium, and helium into heavier elements such as carbon and oxygen. One version of fusion power is used in the hydrogen (fusion, or thermonuclear) bomb. Thermonuclear weapons heat certain light elements, usually the heavy hydrogen isotope deuterium and the light metal lithium, to such high temperatures that nuclear fusion reactions between them can occur. Traces of radioactive hydrogen (tritium, or hydrogen-3) are used to speed up the initial fusion process and boost the gas temperature. The momentary extremely high temperature needed to ignite the fusion reaction is provided by the explosion of a nuclear (fission) bomb.

The use of light isotopes in a fusion reactor has been under experimental study since the 1950s. Since it is inconvenient (and likely to be disturbing to the people in the next apartment) to set off an atomic bomb in your power reactor many times every second to keep the fusing gases hot, electrical power is used to heat the ionized gases hot enough for reaction. For nearly forty years the fusion experimenters have struggled toward the break-even point, beyond which the reactor produces more power than it consumes. After reactor performance improvements of approximately a factor of one trillion, the break-even point is now in sight. The ambitious $8 billion International Thermonuclear Experimental Reactor (ITER), intended to demonstrate the ability to run above the break-even point, will be under construction by late 1998 and will be in operation by 2008. The reactor is widely regarded as the last step before design of commercial fusion reactors. Europe has raised its investment in fusion power to $600 million in 1996, while the total American commitment to fusion-power research has dwindled to $244 million (only a small fraction of which is for ITER). Japan has moved to take up the slack created by America's waning participation. Japan, vulnerable to energy disruptions because of its lack of domestic supplies of oil and coal, has a 1996 fusion research budget of $500 million and is actively seeking to have the experimental reactor located in Japan. Once again the United States, which pioneered fusion research in the 1950s and 1960s, has paved the way for foreign domination of the commercialization of its research.

Since there are many light elements, and several isotopes of most of them, a wide range of choice exists for potential fusion fuels. The reaction that is in many ways the simplest, reacting two ordinary light hydrogen nuclei with each other, happens to be very slow under achievable reactor temperatures and densities. Some of the other candidate reactions are fast enough to be useful, but they give off abundant neutrons. Neutrons are bad news for two reasons: First, since they are uncharged particles, their energy cannot be directly tapped to produce electric power. Second, neutrons escape instantly from the magnetic fields used to contain the electrically charged gases in the reactor, and the neutrons then slam into the metal walls of the reactor. Their energy is not only wasted but squandered making noxious artificial radioactive isotopes, which accumulate and soon present a serious radioactivity hazard.

Kulcinski points out that it would be very desirable to use light helium (helium-3) with heavy hydrogen (deuterium) as the fuel of huge fusion reactors because their reaction produces no neutrons. The energy of the reaction is therefore very easily tapped with high efficiency, and no neutron irradiation of the walls occurs in the main reaction. Deuterium is present in small amounts in water, from which it can be extracted at very modest cost. The amount of deuterium in Earth's oceans is enough to power civilization for millions of years. There are only two barriers to commercial generation of cheap, clean electricity from helium-3 and deuterium. The first is the need to demonstrate operation well above the break-even point in a fusion reactor. We are, as I just mentioned, close to that goal and making steady progress toward it. The second is the lack of helium-3 on Earth. Helium is a light gas that escapes readily from Earth's atmosphere. Any helium-3 that may have been captured by Earth during its formation has had ample time to escape from the planet into space. Actually there is a tiny flow of helium-3 out of Earth's interior, but there is no possibility of capturing this gas before it escapes. Indeed, even if we could capture it, the amount is too small to be a significant source of power. The largest accessible source of helium-3 on Earth is from the decay of tritium in hydrogen bombs. To extract the helium-3 requires melting down every warhead on Earth (an act that is appealing for other reasons), but even so, that would yield only 0.3 tons of helium-3, about enough for a single day of power generation by the United States alone. Increasing the helium-3 supply would mean scaling up the production of the dangerous radioactive isotope tritium.

Kulcinski argues that the solar wind has implanted great quantities of hydrogen and helium into the lunar regolith. The concentration of helium-3 is highest in mature regolith, and within the regolith, highest in the mineral ilmenite. The concentration of helium-3 in the regolith is actually astonishingly small (one ton of helium-3 for every one hundred million tons of regolith), but the total amount on the Moon appears sufficient to power Earth for several centuries. On Earth, no mineral resource with such a low concentration has ever been mined for its own sake. But helium-3 is exquisitely valuable because of its energy content. A good point of comparison is platinum, a well-known commodity of high value, with a price of $20,000 per kilogram, or $20,000,000 per metric ton. At present, the average American pays about ten cents for a kilowatt-hour of electricity. Of this, only about one cent is the cost of fuel. The rest, the "overhead," covers the amortized cost of the generation and distribution equipment, maintenance, salaries, and profit. As our crude oil reserves dwindle and environmental factors (such as a "carbon dioxide emission tax") become more pressing, the fuel cost is certain to rise, roughly in parallel with the decreasing per-capita usage of electric power made possible by improved energy efficiency in power generation, industrial use, home heating and cooling, and transportation. Thirty years hence the price of power is likely to be close to 20 cents per kilowatt-hour, about half due to fuel costs and half due to the "overhead." Your electric bill will not necessarily be larger, because per-capita energy use may then be only half what it is now. The present value of helium-3 fuel is about $1,500,000,000 per metric ton (but, counting overhead, you will pay about $15,000,000,000 for that power). Thirty years from now the fuel itself will be worth about $15,000,000,000 per metric ton (that's in 1998 dollars: inflation will make the numbers look larger). Here is a commodity worth shipping from almost anywhere! Even with shipping costs of a billion dollars per ton, helium-3 would still be a great bargain.

The central question regarding the practicality of extracting helium-3 from the lunar regolith and returning it to Earth is therefore not one of how much it costs to ship it back. The real problem lies in how much equipment and energy are required to mine, extract, separate, and package the lunar helium-3.

The lunar helium-3 scheme exists in several versions, but we need to be familiar only with the basics. First, the story goes, we must transport to the Moon pieces of heavy equipment to dig the regolith, crush the agglutinates, beneficiate the ilmenite, sieve the ilmenite to enrich the smallest grains, and heat those grains to release the trace of solar-wind gases in them. These gases are dominated by hydrogen, nitrogen, carbon, oxygen (made from ilmenite by heating it with solar-wind carbon), and normal helium (helium-4). Of this complex gas mixture, only 0.01 percent is helium-3. These gases must then be separated to get a pure helium extract. Then the helium isotopes must be separated by clever low-temperature processing. The cost of transporting to the Moon the huge mass of equipment needed for all these steps and operating it there is a serious problem indeed. By far the greatest mass is in the mining and regolith-baking equipment.

When discussing the extraction of ilmenite for use in oxygen manufacture (chapter 4), I pointed out that the crushing of agglutinate does a rather poor job of producing pure "liberated" ilmenite grains. I further pointed out that very fine particles, which are precisely the ones we want in this case, are extremely "sticky" and difficult to separate from other materials. The inability to separate ilmenite efficiently from other minerals means that we will recover only a fraction of the ilmenite, and that the ilmenite-rich separate will be extensively contaminated by other materials.

Suppose we start with a regolith sample containing 10 percent ilmenite (real lunar regolith contains 0 percent to 20 percent ilmenite locally). We must use rather mature regolith because the concentration of solar-wind gases is highest there; thus, agglutinates must be abundant. We must crush the agglutinates because most of the ilmenite is welded to agglutinate grains and cannot be separated. If we crush the agglutinate coarsely, each fragment will contain glass and two or three other mineral grains, making electromagnetic separation ineffectual. If we crush the agglutinate to a small particle size, then most ilmenite grains will be liberated, but fine dust is extremely difficult to separate electrostatically or magnetically because of the stickiness (really, cosmic "static cling") of the particles. We may hope to end up with an ilmenite-rich separate that has about 8 percent of the mass of the regolith sample and a concentration of about 40 percent ilmenite. In other words, our enriched material is more than half dross. Further, it contains only about a third of the ilmenite in the original sample. Now suppose we begin with 100 tons of regolith, and it takes us one unit of energy to heat each ton to the point at which it releases its solar-wind gases. If we use crushing and beneficiating equipment, we end up with only 8 tons of ilmenite concentrate to heat, containing 3 tons of ilmenite fines. We use 8 units of heat energy to get out 30 percent of the helium-3. On the other hand, if we discard all the crushing, beneficiating, and sizing equipment, and we heat unprocessed regolith, we use 100 units of energy to get out 100 percent of the helium-3 (twelve times as much energy to recover three times as much helium-3). This is a quarter as much helium-3 per unit of energy, but it saves the cost of transporting a vast mass of rock-crushing and sorting equipment to the Moon, a tradeoff that will very likely save money. Rock-crushing and -handling equipment is especially vulnerable to breakdowns in a gritty, dusty vacuum with wide temperature excursions each lunar day. The prospect of servicing and repairing such equipment under lunar conditions is simply frightening.

When all the bookkeeping is done, the "less-efficient" option—without crushing, beneficiating, and sizing—looks more practical. The total energy use to extract the helium-3 and purify it is about a quarter of the energy content of the helium-3, so the scheme can be operated profitably in terms of energy payback. The total energy return is enough to meet all human power needs for several hundred years. This means that the energy investment in the scheme is about enough to power Earth for a century. That is a truly staggering investment (and economic payback is by no means assured even with a 4:1 energy payback).

The only plausible source of the needed process heat for baking out the solar-wind gases is the Sun. Large solar collectors would be deployed, probably on roving processing plants, to focus sunlight onto the baking ovens. These solar collectors must be rather accurately parabolic in shape (i.e., rigid and heavy) in order to focus their energy controllably. Replacing the roving processor and all its equipment with an equivalent mass of solar cells, we would cover a much larger area, perhaps twenty times as large as that of the solar collector. Suppose we have a processor with a thousand-square-meter solar collector. The thermal power delivered is about one megawatt. While operating, it releases a stream of helium-3 with an energy content equivalent to four megawatts. But the replacement of the roving processor with solar-cell arrays would provide us with a *continuous supply* of energy over twenty thousand square meters (at 20 percent efficiency, about four megawatts of electric power), not simply a one-shot gleaning of helium-3. Solar power is continuously available, but the helium-3, once harvested, is gone. There is a limited sense in which helium-3 is a renewable resource, but the helium-3 influx from the Sun would take millions of years to renew the supply up to mineable concentrations. Incidentally, darkness has no effect on this comparison of the helium-3 and solar-power scenarios, since both the solar-heater and solar-voltaic systems shut down at night.

In principle, mature regoliths anywhere in the inner solar system would also be good sources of helium-3. Mercury is unfortunately so deep in the Sun's gravity well that return of material from its surface is very demanding, and it is so hot in the daytime that helium may be baked out of the surface and lost into space. Venus, Earth, and Mars are all shielded from the solar wind by atmospheres and (in part) by magnetic fields that deflect the solar ions before they can strike the planet. Small NEOs are likely to have little regolith and virtually no mature regolith. The satellites of Mars may have mature regoliths, but their surface area is minuscule compared to the Moon's. Between them, Phobos and Deimos should contain at least a thousand times less helium-3 than the Moon, enough to power Earth for a few months at most.

Whatever the fate of future economic analyses of SPSs, Lunar Power Stations, transportation of metals for SPS construction from near-Earth asteroids or the Moon, or lunar helium-3 extraction, it appears that the use of energy from space can relieve our planet of the burdens of fossil-fuel combustion and nuclear waste disposal. I have so far emphasized only the feasibility of deriving energy from novel sources in space. But our future choices of energy systems have a stronger impact on Earth's surface environment than any other technical choice we can make. The environmental impact of humans' future energy demand needs further examination. We return to this topic, with a larger range of technical options at our disposal, in chapter 14.

9

MARS IN IMAGINATION
AND REALITY

The role of the Moon in human history was to serve as the first target against which the emerging technologies of space flight were exercised. The role of Mars was to serve, both before and after the first lunar landing, as the inspiration for space travel.

—Metodius K. Pateyev
Vpered, na Mars!
2077 AD

The motor whirred faintly through the thin Martian air, slowly pulling the drill stem and core out of the borehole. Alli closed her eyes and waited in boredom. Her thoughts drifted back to the exciting time of her arrival at Mars just three months earlier.

Ares 2, the first international mission sent to land on Mars, had coasted silently in low Mars orbit, some six hundred kilometers above the rusted landscape. The crew, still on an emotional high from the orbit insertion burn two hours earlier, clustered by the windows and gaped like tourists at the scenery.

Vast canyons scrolled over the horizon into view. The snow-dusted Tharsis volcanoes to the north and the vast Valles Marineris canyonlands to the east set an intimidating scale to the scene. Ancient landslides of immense extent scarred the slopes of the valley system, spreading fans of debris far out across the canyon floor. As the valley system rolled beneath them, great networks of ancient riverbeds appeared ahead, seeming to drain from the cratered highlands down into the rolling lowlands beyond, the terrain where geologists of the previous millennium had seen many signs of lakes, glaciers, and even shallow seas.

But the water was now long gone; the seas a parched waste of sand and rocks, of salt beds and red clays left by Mars' clotting, drying lifeblood. Vast flood channels—wider, longer, and deeper than the Amazon—stood empty, mute testimony to bygone glory.

Mars, on its face, was a cautionary tale of a planet with hopeful beginnings that had somehow gone horribly wrong.

The motor shut itself off. Alli opened her eyes to see two meters of fresh drill core exposed. She lifted out the core and lugged its mass, more than she could have lifted on Earth, into the windbreak shed and loaded it into the automated logger. The first analytical traces popped up on the GRS screen. Iron, as always, was high and fairly steady over the length of the core. Silica varied little, roughly complementary to iron. Magnesium seemed to be trending upward. Sulfur fluctuated a lot along the core, but it also seemed to be trending upward. Both water and potassium seemed high.

As Alli sat in the morning sunlight on a convenient black rock and ignored its red frosting of fine dust, she watched the monitor screen build up more and more data on the core. She relegated the data collection to background and called up a window to display the relationship of this core to the previous ones from the same hole. No doubt about it— water, potassium, sodium, and sulfur were all increasing as the drill cut deeper into the crust.

Alli quickly chipped a few sample flakes from the core at strategic points and dropped them into the sample loader on the XRD. The X-ray beam started up, and a powder pattern promptly began to emerge and intensify on the XRD screen, revealing the mineral makeup of the first sample flake. The computer tagged the features automatically: gypsum and anhydrite both present; epsomite, several clay minerals, and a complex potassium-calcium sulfate all mixed in with pyroxene, plagioclase feldspar, and a little free silica.

Alli spoke into her suit microphone. "Comp link com one and seven. Gerry, I'm getting more signs of water, salts, and clays in core F36. I think we need to go deeper here before we pack up and move on to site G. In fact, I think a seismo run with a pickup down the hole might be a good idea."

Gerry Sawyer, always pleased at the chance to set off a few explosions, came on the horn. "I'm all suited up and rarin' to go. I can be there in twenty minutes. Why don't you ask comp to dump me the site data you've collected while I'm jogging over."

Alli spoke a few words to set up the file transfer and smiled briefly at the ludicrous image of Gerry actually jogging. Gerry was good at seeming to bustle, but he drew the line well short of actual exercise. Alli sat in the Sun and fantasized about lying on a beach in Hawaii while waiting for Gerry to arrive with his big red backpack of goodies. As she waited, comp finished its analyses, popped up a summary window on the screen, and uploaded a copy of the results to the mission command module in synchronous orbit overhead. She smiled to herself as she thought through the emerging picture. The long-known potassium deficiency of Mars dirt had been a subject for debate since the Viking missions in 1976, almost fifty years earlier. Some hasty scientists had even concluded from this and the feeble atmosphere that cold little Mars was deficient in all volatile materials. That conclusion unfortunately ignored the other striking piece of news from the Viking landers—that sulfur, an element much more volatile than potassium, was enormously abundant in Mars dirt. So those theories were no good—but where was the potassium?

Gerry, who had the annoying habit of whistling and even singing to himself with his suit microphone open, strode onstage to the strains of the Grand March from Aida. He unreeled a string of six small seismic detectors as he approached, tamping five of them carefully into the dirt. He bent over the jacketed drill hole with detector number six, complete with an extra length of cable, in hand. "All your gear out of the hole?" he asked cheerfully.

Alli smiled. "Clean as anything can be on Mars."

Gerry carefully lowered the sixth detector some thirty meters down to the bottom of the hole and poured in a few handfuls of fine dust on top. He then took out of his pack a hammer, a length of pipe, and a fluorescent green cylindrical object that looked like an overgrown firecracker. Walking off, he unraveled a fine wire from the charge, then pounded the pipe an arm's length into the ground and dropped the charge down it. He then carefully pulled the pipe out, still threaded on the wire from the charge, and laid it down nearby. He then walked over to Alli's shelter and invited her inside, out of line-of-sight to the charge. He flipped a little control pad out of his vest pocket and punched a few buttons. "Seismo comp's up and running, and patched into your data comp. All the seismic detectors are uncaged, and the charge is warm. Ten seconds to go."

The explosion sounded even smaller than it was, its sound muffled by a thick pad of soft dirt and carried to them by the very tenuous carbon dioxide atmosphere. The ground briefly twitched beneath their feet. Gerry didn't even look up from his displays. Alli peeked around the windscreen, half expecting to see a shower of rocks falling from the sky, but there was just a narrow, twisting column of dust above the shot hole.

"Hot diggity, good data! OK, the sensors are caged. It's a wrap." Gerry gazed at the plot from the deep sensor and shook his head in wonder. "Incredible! There's nothing like this on Mars—it's like a natural gas dome in Texas! Better uplink this stuff and get some independent opinions." He started to command Alli's comp to send the data stream upstairs, but he halted at a strange sound. The wire to the down-the-hole sensor had pulled taut.

"Weird." Gerry glanced at Alli and they both trotted over to the hole. Gerry knelt on the red dust cushion and tugged at the line. It rose freely, taut with the weight of the sensor. Gerry considered the layout for a moment. "You know, this thing is a good five meters deeper than when I set it in. How deep did you drill?"

"The last core ended at 32.6 meters." Alli frowned inside her helmet. "Can you tell me how much line you have out?"

"Sure, it's color-coded on the wire. Look here—this is the forty-meter mark, almost in the hole. And the whole wire is only fifty meters long."

Gerry disconnected the sensor wire from the instrument package and gently lowered the sensor head into the hole. It stayed taut as it paid out all the way to the fifty-meter mark. "You've got yourself a cave here, Alli, or somebody's mine." He pulled the wire out hand over hand, fussing with it a bit to keep the sensor from jamming in the borehole casing. Finally the detector emerged into the light of the Martian day.

"It's smoking! It must have a short!" Alli shopped in astonishment.

The sensor head was coated with a fuzzy layer of fine white crystals. The whole sensor head was surrounded by a wispy cloud—and it was clearly, obviously wet.

"There's nothing wrong with the sensor, Alli; there's something right about the planet. That's saltwater—brine. You've found your potassium, and I'll bet you've found the missing ocean too."

—River Run Deep

In the late 1800s and early 1900s, the Mars of the astronomer Percival Lowell and the science fiction writer H. G. Wells was a dying planet, its oceans drying up and its atmosphere escaping into space. Massive civil engineering projects, vast networks of canals, carried the seasonal melt water from the fringes of the polar caps to lower latitudes. As befits an ancient planet, it was peopled with an ancient race of superintelligent, technically accomplished, and utterly inhuman aliens, who mastered interplanetary flight in order to have a try at conquering the younger, lusher, and technically backward Earth. The Barsoom of the writer Edgar Rice Burroughs, a surprisingly populous desert, was peopled with many races of sword-wielding aliens, as found by John Carter when he was mysteriously transported there. Even the Mars of mid-twentieth-century writers such as Ray Bradbury comes equipped with the shadowy descendants of ancient races, superior in physical and mental technologies to the upstarts from Earth. Indeed, many recent science fiction writers have invoked hibernating ancient races of Martians, resting in high-tech tombs against the day when the Martian spring may again arrive. The Mars of the imagination is not just the home of alien life; it is an abode of ancient intelligence.

The facts are less encouraging. Mars is very nearly twice the diameter of the Moon and half the diameter of Earth. It is not surprising that its geological evolution also lies between these extremes. The Moon experienced only a brief period of intense internal activity early in its history. Hardly any lunar rocks are younger than three billion years. A small percentage of the lunar surface is covered by dense, dark flows of basaltic lavas, occupying basins blasted through a thick global highland crust of anorthosite by the impacts of comets and asteroids.

No lunar atmosphere or hydrosphere exists, nor has any existed for its entire span of recorded history. The Moon has a very small core—if one at all—that must have formed when the Moon was still warm and active, very early in its history. Earth, in contrast, is still a powerful and active geological machine. Almost two-thirds of Earth's surface is less than two hundred million years old. The silica-rich continental blocks form highlands, covering almost a third of the surface area of the planet, which mostly emerge above sea level. Rocks older than three billion years are quite rare on Earth because continuing geological activity, driven largely by heat from the radioactive decay of uranium, potassium, and thorium, constantly recycles Earth's older crust. Earth has an actively convecting dense metallic core that produces and sustains a strong magnetic field. Many areas of volcanic activity, some of them associated with extensive basaltic lava flows, are found widely distributed over the planet. Earth has both a massive complex of oceans and a substantial and complex atmosphere that together conspire to sustain an extremely varied and dynamic biosphere supporting many millions of species. Life has in turn strongly influenced the present state of Earth's surface, infusing the atmosphere with oxygen and small but important traces of methane, ammonia, and other gases. One indirect consequence of the presence of life on Earth is the recent addition of a variety of distinctive unnatural gases, such as chlorofluorocarbons, to its atmosphere by one of its dominant species.

Middle-sized Mars occupies the middle ground between Earth and Moon in almost every respect. We know that Mars underwent extensive internal melting early in its evolutionary history. Density separation of its melted components produced a dense core, a mantle, and a low-density silicate crust. Gases vented by volcanic activity accumulated to form an atmosphere and a transient hydrosphere. A weak magnetic field was generated by motions in the electrically conducting core. It is even possible that primitive life originated in the ancient lakes and seas, but we do not yet know for sure.

As part of the bicentennial festivities of the United States, two Viking spacecraft were landed on Mars in 1976. They were designed principally to search for direct and indirect evidence of life on the Martian surface. A very sensitive gas-analysis experiment was used to analyze the atmosphere and to test the gases released from heated surface materials to search for traces of organic matter. The results were strongly negative. Many simple organic substances were demonstrably absent from both exposed surface samples and excavated dirt at the instrument's sensitivity limit of one part per billion.

The atmosphere itself, as analyzed by the Viking landers, is a treasure trove of seemingly paradoxical features. The atmosphere is, in most places and at most times, about 96 percent carbon dioxide. The remaining 4 percent is almost entirely a mixture of similar amounts of nitrogen and argon. At this level of detail, the atmosphere of Mars is quite similar to that of Venus. The Martian atmosphere contains small amounts of carbon monoxide and oxygen from solar ultraviolet destruction of both carbon dioxide and water vapor, and small (but highly variable) traces of water vapor. There are also tiny traces of other noble gases besides argon, namely, neon, krypton, and xenon.

The abundances of three argon isotopes tell an interesting story. Argon-36 and argon-38 are "primordial" isotopes that date back to the origin of the solar system. The primordial isotopes of argon and the other noble gases are released into the atmosphere the first time the rocks containing them melt. Argon-40 is "radiogenic," meaning that it is produced by radioactive decay of potassium-40, a minor isotope of potassium, a ubiquitous constituent of planetary crustal rocks. The amount of argon-40 on a planet at the time of its formation is extremely small, but as billions of years pass, argon-40 is made abundantly by radioactive decay. The *rate* of formation of this radiogenic argon-40 is highest at the time of planetary formation, and it drops off rapidly thereafter: the present amount of potassium-40, and hence the present rate of formation of argon-40, is only a tenth of what it was when the planets were young. If a planet remains very cold, the radiogenic argon remains trapped inside the potassium-bearing minerals in which it is formed. But if there is continuing geological activity on the planet, the melting of rocks releases the gases contained in them (notably argon-40 from potassium-40 decay and helium-4 from uranium and thorium decay). These gases escape into the atmosphere during volcanic eruptions. Helium, however, is so light a gas that it escapes readily from the top of the atmosphere into space. Argon-40, too heavy to escape, accumulates in the atmosphere. On a small planet like Mars we expect the rate of outgassing to decrease rapidly with time, as the radioactive elements die out and the planet cools down. We can tell what the original abundances of the radioactive isotopes were by examining very ancient solar system materials (unmelted meteorites from primitive asteroids), and we can measure the decay rate of these various isotopes in the laboratory. Then we can measure the abundances of the radioactive elements and their decay products in natural samples (differentiated meteorites, Martian meteorites, lunar and terrestrial rocks) and calculate their ages. Every radioactive isotope provides an independent clock for measuring geological time. Usually when we date ancient rocks, we can compare the readings of two, four, even five or six of these independent clocks and find excellent agreement among them. Any evident discrepancies are almost always due to reheating of the rock, causing partial loss of the gases produced from decay of potassium, uranium, and helium. Fortunately, we have other dating systems (uranium-lead, lead-lead, rubidium-strontium, rhenium-osmium, and so on) that do not produce gases and are therefore quite stable and reliable clocks.

The abundance of the primordial argon isotopes in Mars's atmosphere is so small that, according to some theorists, only a small fraction of the mass of the planet could ever have melted and released its burden of primordial gases. But this interpretation contradicts the geological and geophysical evidence showing clearly that Mars is a thoroughly melted and differentiated planet. Meanwhile, the low absolute abundance of radiogenic argon-40 suggests a poorly differentiated planet that has not undergone extensive melting, separation into layers, and gas release to the atmosphere. But the abundance of argon-40 *relative to argon-36 and argon-38* is much higher than on Earth, suggesting that Mars outgassed *relatively later than Earth,* when the argon-40 abundance had had sufficient time to accumulate in quantity. Since we know Earth is still outgassing vigorously today, the implication that Mars outgassed relatively later is ridiculous. Fortunately, a way out of this apparent paradox exists. The easiest explanation seems to be that almost all of the atmosphere of Mars released in its earliest few billion years of evolution has been lost from the planet. Only the most recently released 1 percent of the gases remain today. It certainly would be interesting to compare the argon-40 content of the atmosphere to that of a range of Martian crustal rocks, to see whether recent outgassing (causally connected with recent volcanism and mountain building) has in fact been important.

We have not retrieved documented samples from the surface of Mars and subjected them to radioactive dating in laboratories on Earth. But, by a wonderful stroke of fortune, nature has provided several free samples of the Martian crust in the form of meteorites. These meteorites belong to four different composition types, all rare classes of achondrites, and all bearing the distinctive chemical imprint of gases adsorbed from the Martian atmosphere. Dating of several of these meteorites has shown rock ages as short as 1.3 billion years, far less than the 4.6 billion typical of asteroidal meteorites. We can hardly expect three or four random impacts to have sampled the very youngest rocks on Mars. There is some reason for cautious optimism that geological activity on Mars, which certainly persisted for at least 3.3 billion years, may actually continue to the present. If so, then ore formation may have persisted over the entire history of Mars and may even still be going on today. Presently active areas might have abundant hydrothermal (hot-water) activity, perhaps steam venting, or even hot springs and geysers. Both the water and the heat from such active areas would be of great value to a Mars expedition. But hydrothermal areas on Mars may also be, like the ocean-floor hydrothermal vents on Earth, refugia in which ancient forms of life could not only survive but diversify and prosper. (Readers interested in this "alien environment on Earth" will find an interesting account in the book *Deep-Ocean Journeys*, by Cindy Lee Van Dover.)*

Mars today is but a ghost of the planet it once was. The tenuous atmosphere, less than a hundredth of Earth's, weakly protects a desolate, rust-red surface of rocks and dunes. Cosmic radiation readily reaches the surface through the feeble resistance of a tiny magnetic field and the wisp of carbon dioxide gas that passes for Martian air. Small or fragile meteorites, which on Earth would disintegrate or burn up at high altitudes, penetrate through the atmosphere to blast the surface with high-velocity impacts. No liquid water is found anywhere on Mars's surface. In the polar regions, extensive seasonal coatings of dry ice (solid carbon dioxide) frost the landscape, in places accumu-

*Addison-Wesley, Reading, MA, 1996.

lating atop thick layers of water ice. Killing ultraviolet sunlight penetrates to the surface, tearing apart molecules of water vapor and carbon dioxide, making deadly oxygen atoms and molecules of ozone. Organic matter on the Martian surface would be literally burned away by these devastatingly powerful oxidizers. Thus, the absence of organic matter in the Viking analyses makes sense, whether or not Mars ever had life.

Explorers approaching Mars will see that the surface is widely and deeply scored by the valleys of ancient rivers comparable in flow to the Amazon. About half of the planet, mostly in the southern hemisphere, is a Moon-like, heavily cratered highland, but with the terrain softened by wind erosion. The northern hemisphere is a much more Earth-like montage of rolling lowlands, great volcanoes, and features that bear the imprint of ancient water. Geological features ground out by ancient glaciers are seen overlaid by the scars of recent wind erosion. The shorelines of ancient seas can be traced uncertainly along the margins of the great northern lowlands. Impact craters have great dune fields marching across their floors. A few gigantic volcanoes rear up from the surface, their gently sloping, lava-scored cones reaching up and up to as high as twenty-six kilometers above the plains. Their cratered summits almost emerge from the top of the atmosphere. But the volcanoes are quiet now, perhaps completely dead. Great mountains are no longer being built on the surface of Mars. The endless grinding down by windblown dust continues; the blasting of the surface by impacting comets and asteroids continues; the violent temperature extremes that crack rock and the chemical erosion that turns hard minerals to fine dust continue. Mars continues, but it is nearly dead—or is it merely sleeping, awaiting the kiss of life to revive it to wakefulness?

The craters made in the highlands by large comet and asteroid impacts give strong evidence of a deep layer of permafrost and ground ice over most of the surface of the planet. At latitudes greater than about thirty degrees, impact craters have a very peculiar appearance never seen on the Moon. They look, instead, like craters made by explosions in mud or wet sand, surrounded by aprons of ejected muck. Most of Mars seems to be covered with a layer of permafrost, kilometers deep in places.

The seas and rivers, lakes and glaciers, are gone; the dynamic play of rain and liquid water is absent. Impacts have eroded away the once Earth-like atmosphere to a feeble remnant, powerless to blanket the surface against ultraviolet rays, to heat it in the daytime, or to keep it warm at night. Much water has been lost forever from Mars, blasted into space by comet and asteroid impacts. Even the water brought in by comet impacts is lost from the planet with high efficiency, since the fireballs from these high-energy explosions easily punch their way out through the tenuous atmosphere, accelerating into space and leaving the planet forever (sometimes accompanied by chunks of rock blasted out of the impact crater, which may later land on Earth and become known as Martian meteorites). More water has been destroyed by ultraviolet sunlight, torn apart into oxygen and hydrogen. Hydrogen flows off at once into space, little affected by the weak Martian gravity, while the oxygen rusts away iron-bearing minerals in the crust, helping to wear down the land and blanket the surface with fine red dust. But much water remains in the vast permafrost deposits, in polar ice caps, and in clays and hydrated salts—immobilized, bound in lifeless minerals. The water is imprisoned, not gone.

There is no *liquid* water on the surface of Mars today because the pressure of the atmosphere is very low. If water, even ice water, were exposed to the low pressures on the surface, the water would quickly boil away. The image is astonishing: boiling ice water! But that is indeed what happens when water is exposed to the low pressures found on Mars. The external pressure on a water droplet (the atmospheric pressure) must be higher than the internal vapor pressure of the liquid to keep it from boiling. To prevent ice water from boiling away on the Martian surface, the total atmospheric pressure must be raised. (This can to some degree be accomplished by descending into the lowest basins in the Martian crust, where the air pressure is nearly twice as high as average.) Since the atmosphere is mostly carbon dioxide, it would be best if we could find some source of additional carbon dioxide on Mars that might be transferred into the atmosphere, some storage reservoir for additional atmosphere. In fact, a considerable amount of carbon dioxide is tied up on the surface as dry ice at the winter pole and is adsorbed in the surface regolith. The best way to get that carbon dioxide into the atmosphere is to warm the entire planet.

That would drive most of the adsorbed gases out of the surface dirt and evaporate the polar dry ice.

But why is the temperature of Mars so low that this does not happen spontaneously? The answer is that the greenhouse effect is very weak on Mars. The greenhouse effect is the warming effect of an atmosphere upon a planetary surface, caused by the ability of certain gas mixtures to transmit visible sunlight efficiently to the planetary surface, while absorbing infrared (heat) radiation from the surface efficiently. The thin carbon dioxide atmosphere on Mars admits sunlight to the surface with very little attenuation. Sunlight is absorbed by dark surface materials and heats the surface to temperatures that sometimes approximate normal room temperature on Earth. Infrared (heat) radiation emitted by the surface is partly absorbed by the carbon dioxide atmosphere, but wide wavelength intervals in the infrared can pass through carbon dioxide gas without absorption. The surface therefore can cool rapidly by dumping its radiation into space. Temperatures plummet to minus sixty degrees or lower at night even at the equator in midsummer. If the mix of atmospheric gases can absorb most of the infrared radiation, then the surface cannot radiate away its heat efficiently, so it heats up. This is the well-known greenhouse effect.

The greenhouse effect operates much more effectively on Earth, sealing in the infrared radiation, raising the daytime temperature, and greatly reducing the radiative cooling at night. The reason for the much stronger greenhouse effect is simply that water vapor and carbon dioxide have complementary absorption bands in the infrared: wherever one transmits infrared radiation, the other absorbs it. Earth has only 0.3 millibars of carbon dioxide and Mars an average of about 6 millibars. The failure of the Martian greenhouse effect is obviously not due to any shortage of carbon dioxide! The problem lies in the virtual absence of water in the Martian atmosphere. Water fails to enter the Martian atmosphere because the surface temperature is very low. The vapor pressure of water and ice varies roughly exponentially with temperature (each temperature decrease of five Celsius degrees lowers the vapor pressure by about a factor of two), so the low nighttime and polar temperatures dry out the atmosphere very effectively, just as in the polar regions of Earth. In the daytime near the equator, the immediate surface is much warmer than the regolith is at a depth of a few centimeters. Any water vapor introduced at the surface will rapidly diffuse down to depths having a low, nearly constant temperature that reflects the average of the daytime and nighttime temperatures. The vapor will immediately condense there in that "cold trap," adding to the permafrost layer or simply being adsorbed on the surfaces of fine dust grains.

The chain of logic is impeccable. We asked why Mars is so cold, and we have traced the explanation to its root. It is cold because of the failure of the greenhouse effect caused by the cold-trapping of water vapor on and in the surface. Mars is so cold because it is so cold!

10

HOMESTEADING
MARS

*The rising Sun swept across the solar-cell assembly atop the Coradinis'
little dome. It creaked and popped under the thermal stress, dispelling a
faint dusting of dry-ice frost. Pietro and Elisabetta rose to face the chill
Martian dawn. Serious and intent, they folded up their bunk and set efficiently about their daily routine. They had a lot on their minds: this
was their big day. If this didn't pan out, all the bootlegged time and
equipment would be wasted.*

*The output from the solar-cell array stabilized while they ate a hasty,
forgettable breakfast of dried this, preserved that, reconstituted something, and recycled water—appropriate enough fare to accompany mechanically recycled, stale air. Their thoughts ran in the same channels,
but they said little as they ate.*

*Pietro cautiously edged around the tiny, fold-down table, cleared the
dirty dishes into the tiny fold-down sink, rinsed them in a trickle of recycled water, and folded the table and sink away. He stood in the airlock as Elisabetta methodically pulled on her suit, faintly redolent of
thousands of hours of use. As soon as she had the suit in place, she
changed places with Pietro, without any word or signal, as he donned
his suit. In such cramped quarters, it was easier to help each other with
their bulky helmets and backpacks than to manage them alone. The familiar ritual of suit check occupied the next few minutes.*

*They took turns cycling through the narrow airlock, each bulky suit
nearly filling the modest volume of the lock. Outside the outer airlock*

door, the view was suddenly vast and uncluttered. Each stretched and took a deep breath, looking away from the dome at endless red vistas, broken only by the partial profile of their equipment dome, nestled nearby in a hollow, and already stained a native red by Martian dustfall. By design, none of the rest of the base complex was visible from their front door. It was uncomfortable even to look at the tiny dome from which they had just emerged.

Pietro stumbled a bit as he stretched and took several exaggerated steps to work the kinks out of his muscles, easily catching and righting himself in the weak Martian gravity. Both of the Coradinis still felt awkward in gravity after years spent mostly in transit and on asteroids.

Their equipment dome stood unpressurized, the outer door closed to minimize intrusion of the ubiquitous fine, red dust. Pietro spun the latch and pulled the outer airlock door open. They filed in through the open inner door. Elisabetta pulled the outer door shut and dogged it tight, then threw the main power breaker for the dome. The computer, kept from violent environmental extremes by its insulated case and battery-powered heaters, switched itself to external power. Seriously, intently, they ran through a forty-item checklist as they powered up the mechanical equipment and processing modules.

"OK, Pietro, let's go for it. Everything is green."

Pietro switched on the power for the submersible pump at the bottom of the borehole, deep within the brine-filled cavern far beneath them. The preserved lifeblood of Mars still flowed sluggishly through deep arteries: here lay the shallowest known brine aquifer on Mars.

The telltale lights showed the pump operating smoothly, forcing a column of cold, dense brine up the borehole pipe. Within moments the brine began gushing out into the first processing tank. Heaters drove off the water as a tenuous vapor, to be condensed in another part of the apparatus. A variety of sensors monitored the composition and quantity of the water as it accumulated.

Some of the brine was drawn off into an electrolysis cell, where a low-voltage DC current separated the water into hydrogen and oxygen. The hydrogen stream from the cathode was bubbled through a tank of rusty Mars dust heated to a dull red heat. That reaction produced metallic iron and gave off water vapor, which was condensed and saved in a tank. The oxygen stream from the electrolysis cell vented through a series of pressure regulators and mixing valves directly into the dome, slowly filling it with Martian oxygen.

Nitrates from the brine were separated from sulfates, chlorides, phosphates, and carbonates by an ion-exchange column. The nitrates were accumulated in a separate reactor that extracted nitrogen oxides from the nitrates and catalytically split them into nitrogen and oxygen. Both gases were vented into the dome.

The pressure gauge now registered a quarter of an atmosphere. The mix was about half oxygen and half nitrogen, with the original trace of carbon dioxide that had filled the dome still present. The tension continued to mount in silence until oxygen reached a pressure of 0.2 atmospheres. Waste heat from all the equipment had raised the internal temperature of the dome well above the freezing point.

Pietro pointed at the pressure gauge. "OK, this is it."

They ceremonially removed their helmets and took a breath of real, uncycled Martian air—air from deep inside Mars, stored as salts and minerals to await the next Martian spring. The smell was a little strange, fresher than they were accustomed to, and free of old food and body odors, with a faint tang that reminded them of thunderstorms on Earth. It was incredibly invigorating.

Elisabetta carefully tapped a cupful of water from the water tank into a battered geological sample jar. She raised it to Pietro's lips, and he drank. Deeply moved, he solemnly repeated the act for her, unconsciously imitating the communion ritual of his youth back in Tuscany. The water was cold, clear, and almost tasteless—water from deep inside Mars, long stored against the day when life should again appear to drink it.

Elisabetta giggled briefly to herself, suppressing her mirth as if she were in a sacred place. Pietro looked at her in surprise, then burst out laughing. Within moments they were both laughing uproariously, hard enough to bring tears to their eyes.

Pietro gestured broadly toward the display screen on the gas chromatograph that monitored the output of the nitrate decomposer. "Look—look at this!" he gasped through his laughter. Elisabetta stepped over to the GC and read the chart. She immediately resumed laughing, holding her aching sides through the bulky Mars suit.

"Hey, Pietro," she gasped, "I think we have a bug to work out in our *system." She sagged weakly to the floor, still laughing. "Look, nitrous oxide . . . "*

Pietro dogged down his helmet and helped Elisabetta into hers. They sat side by side on the floor for several minutes, breathing stale recycled suit air as they got back to normal.

"Great water," said Pietro.

"Great air," said Elisabetta.

"Great laughing gas!" they chorused, again dissolving into hysteria.

—Deep Breath

The National Aeronautics and Space Administration was born as a can-do agency. In 1961, President John F. Kennedy charged NASA with the staggering task of executing manned expeditions to the Moon within ten years. NASA responded by doing the job in eight years. In 1964, during the exciting morning of the Apollo program, NASA knew it had a bright future reaching far beyond Apollo. Projecting beyond the first manned landings on the Moon, NASA planned a rapid transition to the construction and use of a lunar base in the 1970s. By 1980, in NASA's view, the first manned expedition could be dispatched to Mars on nuclear-powered rockets. *A NASA that had not yet sent astronauts to the Moon was confident that it could land an expedition on Mars within sixteen years.* A permanent Mars base could be a reality by the late 1980s. Manned exploration of the satellites of Jupiter could be expected by the end of the century—by 2001. Arthur C. Clarke's ambitious mission to Jupiter in 2001 was no flight of fancy. In fact, it was solidly based on what NASA itself thought it would then be doing.

Since the early 1980s, NASA has sung a different tune. When asked how we would go about returning to the Moon, NASA dusts off a scheme that would permit the first lunar landing "only" sixteen years after the go-ahead. And of course NASA has no go-ahead. *We are twice as far from the Moon now as we were in 1961.* How the mighty have fallen! There is, in simple fact, no longer any NASA schedule for a return to the Moon. The manned lunar base is permanently over the horizon. Manned expeditions to Mars, though warmly endorsed by President George Bush, never even made it to Congress. The reason is simple: NASA has failed to make space as accessible to us now as it was in 1961. The cost of spaceflight has risen, not fallen, with experience. NASA has been run as an entitlement program, not as an aggressive innovator and pathfinder. The National Aeronautics and Space Administration has focused more of its energies on the final A in its name than on the preceding A and S. While some supporters of space exploration see congressional approval of a manned expedition to Mars as the key to a renewal of NASA, others, including the present Administrator of NASA, Daniel Goldin, see the renewal of NASA as an essential prerequisite to winning political support for advanced exploratory missions.

My purpose here is not to advocate a carte blanche for vastly overpriced NASA programs. It is rather to explore how common sense and uncommon technology can bring down the costs and raise the effectiveness of space exploration, especially the exploration of Mars. My ultimate purpose, here as elsewhere in this book, is to examine the feasibility of making space enterprises self-sufficient.

I began the discussion of the use of lunar materials by asking what needs of an early lunar landing could most readily be met by using local materials. I then attempted to assess how much advantage accrued from each such possible use. I shall adopt the same approach for Mars. However, the exploration of Mars is in a somewhat more primitive state than that of the Moon. Automated and wheeled roving landers, roving "hoppers," and automatic sample returns all logically precede a manned Mars landing mission, but these have not yet been accomplished.

∴ ⁂

The most obvious way to enhance long-range rovers and "hoppers" (and sample-return missions) is to provide them with propellants derived from Martian materials. To do this, we need an easily accessible source of volatile materials. The most obvious source, and one that is ubiquitous on the Martian surface, is carbon dioxide in the atmosphere. The high-temperature electrolysis scheme that we explored in connection with a lunar base can separate carbon dioxide into oxygen and carbon monoxide. Run in reverse, acting as a fuel cell, the same apparatus can "burn" carbon monoxide and oxygen in a controlled manner to generate carbon dioxide and electrical power.

A sample-return mission to Mars follows a flight profile similar to those discussed for missions to the Moon (chapter 4) with a few important differences. Beginning with departure from the space station in low Earth orbit (LEO), the delta V required to fly to Mars is slightly larger than that required to fly by the Moon or enter highly eccentric Earth orbit (HEEO). The second engine burn occurs upon arrival in the vicinity of Mars. It is then necessary to match Mars's orbit about the Sun and descend to a controlled landing, using aerobraking and a brief rocket firing for the last phase of descent. The orbit matching and deceleration may be performed in several different ways, including doing it all by aerobraking. Another option features aerocapture into a highly eccentric Mars orbit (HEMO), a small propulsive orbit-trim maneuver to lift the low point of the orbit (periapsis) out of the atmosphere, a prolonged orbital survey phase (with deployment of one or more communications relay satellites for later use by the lander when Earth is below the horizon), and then a final dive into the atmosphere for entry and landing. A final option, noteworthy for its high propellant usage, would be to use engine firings to perform the entire capture and landing sequence, without reliance on aerobraking. For our purposes, we shall assume the first of these options, aerobraking to landing upon arrival at Mars.

The presence of an atmosphere on Mars, which permits aerocapture and saves great quantities of propellants, makes the job easy for the rocket motors that carry the spacecraft to Mars. The delta V requirement for flight from LEO to the surface of Mars using aerocapture is only 4.8 kilometers per second, compared to 6.1 kilometers per second to land on the Moon. When allowance is made for the mass of the heat shield that must be carried to Mars for the aerocapture maneuver, a given-sized rocket departing from LEO can still deliver about the same mass of payload to the surface of the vastly more distant Mars that it can to the Moon!

Once the vehicle has landed safely on the Martian surface, it may dispatch small rovers to explore the vicinity of the landing site, select samples uncontaminated by the landing vehicle engines, package these samples, and bring them back to the lander. Such local rovers, which need only limited range, could be powered by batteries.

Once the samples have been retrieved and stored, the vehicle blasts off from the Martian surface and accelerates to a high enough velocity for the return trip to Earth. Earth arrival is accomplished by aerocapture, followed by descent of the sample container under a parachute. This scenario is especially easy to envision because it is exactly the way the Soviet Luna spacecraft retrieved samples from the Moon in the 1970s.

Although sending a payload to soft-land on Mars is closely comparable to sending the same-sized payload to a lunar base, the return trip is quite another matter. The delta V for the vehicle to depart from the surface of Mars and enter a trajectory that intersects Earth (for aerocapture in Earth's atmosphere) is 7.8 kilometers per second, compared to only 3.0 kilometers per second for an aerobraked return from the Moon. The amount of propellants required to return from Mars is several times as large as for a return from the Moon. If, as in the Luna missions, the propellants needed for the return trip are all carried from Earth, then the mass that must be landed on Mars is increased several-fold. Thus, the mass that must be sent from Earth to Mars in the first place also has to be several times larger, which drives up launch costs from Earth. Because the Mars return mission is so demanding, the benefit conferred by using Mars-derived propellants is great.

There seems little doubt that the essential "foot-in-the-door" technology for Mars missions is carbon dioxide cracking into carbon monoxide and oxygen, followed by liquefaction of the products. This technique was proposed twenty years ago by Robert Ash, Giulio Varsi, Warren Dowler, and Kumar Ramohalli at the Jet Propulsion Laboratory. Continuing research on this scheme, Kumar Ramohalli, now of the University of Arizona, and his team have recently shown that a small mass of equipment, landed anywhere on the planet, can produce enough propellant for the return trip to Earth. Further, they have built and tested devices that crack carbon dioxide into carbon monoxide and oxygen with high efficiency. The principal requirement for in situ propellant production on Mars is not, as on the Moon, heavy and failure-prone dirt-moving, rock-crushing, beneficiation, and screening equipment. Rather, it is a power source sufficient for the needs of the propellant factory. The two sources of power that first come to mind are solar and nuclear.

A carbon dioxide electrolysis unit requires both thermal and electrical power; indeed, it needs more thermal than electrical power. It is easy to use electrical power to make thermal power, but electrical power is much more expensive in terms of equipment mass, complexity, and reliability than thermal power. Thus, the least desirable approach is to use solar electric or nuclear electric power to meet all the needs of the processing equipment. Instead, it is easy to imagine a simple solar collector, a parabolic mirror, that focuses concentrated sunlight on the processing unit to supply the heat that maintains it at operating temperatures. The main drawback of this approach is that it can operate only when the Sun is above the horizon. Only in the summer polar regions of Mars, just as on Earth, is the Sun continuously above the horizon for weeks at a time. Elsewhere, day and night alternate in a 24.5-hour cycle, again in a manner very similar to the day–night cycle on Earth. Also, because of Mars' greater distance from the Sun, the intensity of sunlight on the Martian surface is about half what it is at Earth's surface. Thus, at almost any plausible landing site *except* the poles, the equipment will lack both solar-electric and solar-thermal power about half of the time and will collect only half of the power the same unit would on Earth while the Sun is shining. To permit the plant to run continuously, electrical storage sufficient to last about thirteen hours is required. (This is vastly easier than storing fourteen *days* of power on the Moon to get through the lunar night!) If this were only the electrical power needed to run the electrolysis equipment, the requirement would not be onerous. But the Sun also provides the daytime thermal power to keep the electrolysis unit at operating temperature. At night, that thermal energy would have to be provided by the only power source available, electricity from the storage batteries.

But what of the great dust storms on Mars? Would not these storms coat solar collectors and solar-cell panels with a thick layer of dust, shutting them down? For several reasons, this does not seem to be a great problem. First, severe dust storms are rare. Second, the solar panels would in any case have to be steerable to follow the Sun across the sky. They could easily be tilted on edge or even upside down during times of heavy dustfall. Finally, a small jet of compressed gas (driven by boiloff from the liquid carbon monoxide or oxygen tank) could be used to blow away any dust that might stick to the critical surfaces. There are no compelling reasons why solar-electric power should not be possible on Mars, given intelligent planning to deal with these problems.

Supplying the electrical needs of the plant from nuclear power alleviates some of these problems. A small reactor or radioisotope thermoelectric generator (RTG) makes electrical power available 24.5 hours per day on Mars. The size of the power system is fixed by the need to provide all the processing plant power needs, both electrical and thermal, from its electrical output.

Recently, because of political winds of change associated with the disintegration of the Soviet Union, another interesting option has become available. It is now possible to buy pieces (near critical mass) of high-grade plutonium, originally manufactured for use in nuclear weapons, to use as a powerful thermal energy source. At present, controls on the sale of fissionable materials from Russia, Ukraine, Belarus, and Kazakhstan appear to be very effective, so that the chances of plutonium falling into the hands of weapons merchants or terrorist groups are slight. Still, the Russian press has exposed several daring plots to steal and export fissionable materials, all of which were frustrated just in time to prevent actual loss of plutonium. The dismantling of the KGB (Committee on State Security) and the dramatic reduction of the power of the military security apparatus (GRU) have been done with a commitment to maintaining the security of nuclear materials. The amusing result is an eager competition between these nations to sell fissionables to "politically reliable" customers such as the United States government! Simultaneously, the demand for plutonium to build American nuclear weapons has vanished, freeing up a large domestic supply as well. NASA missions that would benefit from carrying, say, a kilogram of plutonium (a sphere one inch in radius) would have no difficulty buying it on the international market. Such a piece of plutonium can maintain high temperatures without any external power supply, controls, or monitoring for many years. I must confess that I find it hard to imagine a more satisfying use of weapons-grade plutonium than helping a peaceful international exploratory mission refuel for its return from Mars!

The initial production of propellants on Mars, insofar as we can presently imagine it, would use a nuclear reactor or a solar electric array with battery backup for electrolysis and a plutonium pellet as its source of process heat.

The propellant combination made by this process is sufficient in its performance to return a payload to Earth, but carbon monoxide–oxygen is by no means the best or only conceivable propellant combination. We would never think of using liquid carbon monoxide as a rocket fuel on Earth because it is much more expensive than hydrocarbons and burns with lower specific impulse. Further, carbon monoxide is so toxic that its boiloff vapors cannot be vented into the atmosphere without serious threat to the lives of the ground crew servicing the rocket. Likewise, on Mars carbon monoxide is easy to make but is not necessarily the optimum fuel. It would be doubly beneficial, for example, to make a fuel and an oxidizer that do not require deep refrigeration. Carbon monoxide and oxygen are both gases at normal Mars-surface temperatures and pressures. To liquefy them for storage in propellant tanks, a considerable amount of electrical energy must be expended to run refrigeration equipment. Propellants that boil at a temperature well below the local environmental temperature, and therefore require active refrigeration for long-term storage, are called cryogens. But many fuels, and even some oxidizers, would be liquids under Martian conditions. Since they can be stored indefinitely under local conditions without refrigeration, they are called storable propellants. Are there any storable fuel-oxidizer combinations that might reasonably be made out of the Martian atmosphere by an autonomous propellant plant?

A second motivation for exploring other Mars-derived propellants might be the need for high-performance rocket propulsion for difficult (high delta V) missions originating on the Martian surface. In that case, we would think first of burning hydrocarbons with liquid oxygen, or even using the hydrogen-oxygen propellant combination. But liquid oxygen is a cryogen, and liquid hydrogen boils at about twenty degrees above absolute zero, far below the mean temperature of Mars.

Liquid hydrogen is sometimes called a deep cryogen in recognition of the difficulty of keeping it in the liquid state. Indeed, only one known substance is harder to liquefy than hydrogen: helium gas. Storing liquid hydrogen on Mars—or for that matter, anywhere in the terrestrial planet region—is a difficult, expensive task requiring active refrigeration. Thus, the performance advantages of hydrogen-oxygen rocket engines is bought at a double price: the cost of mining and processing water, and the cost of storing liquid hydrogen.

Since about 96 percent of the Martian atmosphere is carbon dioxide, we must first ask whether there are any other propellants besides the familiar carbon monoxide and oxygen that can be made from pure carbon dioxide. There are in fact two other substances that might be made—solid carbon and ozone. As mentioned with regard to the Moon (chapter 4) gaseous carbon monoxide can be passed through a compact chemical reactor that uses heat and a specially selected chemical facilitator called a catalyst that transforms two carbon monoxide molecules into one carbon dioxide molecule and one atom of solid carbon. Burning this solid carbon with oxygen would release far more energy than burning carbon monoxide, but there is no obvious way to build a rocket engine that uses this carbon powder. One cannot pump powdered carbon like a liquid fuel, nor can the carbon be cast into a tough hollow cylinder, as rubber is in solid and hybrid rocket engines, to be burned by passing oxygen through the center.

Ozone has been considered as a candidate rocket propellant for over sixty years. The ozone molecule, composed of three oxygen atoms (O_3), can be made by the action of ultraviolet light or electrical discharges on normal oxygen gas (O_2). In effect, much of the energy used to make ozone from oxygen is stored in its unstable structure. Liquid ozone is therefore a powerful oxidizing agent that performs better as a propellant than liquid oxygen. However, during German rocket-motor tests in the 1930s, liquid ozone was found to have a most undesirable property that renders it unfit for use as a rocket propellant: it is extremely sensitive to mechanical shock and to vibration. When agitated, liquid ozone reorganizes into oxygen molecules with the release of sufficient heat to vaporize all the liquid. The result is a devastating detonation that completely destroys the rocket. In recent years, ozone has been abandoned as too treacherous to use.

The conclusion from these facts is clear: As long as we are limited to the use of carbon dioxide as the sole source of Martian propellants, we cannot hope to do better than carbon monoxide and oxygen. But what of other constituents of the atmosphere? About 4 percent of the atmosphere consists of argon (mostly argon-40, for reasons we have already discussed) and nitrogen. Argon is useless in chemical rockets because it does not burn or participate in any energy-releasing chemical reactions. Nitrogen, fortunately, is another story.

On Earth, nitrogen is an important ingredient of both storable rocket fuels and oxidizers. The best example of both is the Titan rocket, built by the aerospace division of Martin Marietta (now part of Lockheed-Martin). Originally designed as an intercontinental ballistic missile (ICBM) with a multimegaton warhead, the Titan was used as the launch vehicle for the Gemini manned spaceflight program that trained the Apollo astronauts in orbital-rendezvous and space-walking techniques in preparation for their voyages to the Moon. More recently, Titan 1 and Titan 2 ICBMs have been retired from their role as weapons. A family of satellite-launching vehicles has grown up based on various versions of the Titan.

The Titan was developed because of a conceptual weakness in the first-generation American ICBM, the Atlas. The engines of the Atlas burned a modified aviation fuel, similar to kerosene, with liquid oxygen. This propellant combination performs well and permits a fairly compact vehicle design. Kerosene is obviously Earth-storable, but liquid oxygen (LOX) is a cryogen that is constantly boiling away. Thus, maintaining a fleet of Atlas ICBMs in a state of launch readiness requires either constant attention to the LOX level in the tanks or last-minute LOX fueling from large reserve tanks. The Titan was designed to use only storable propellants, at the price of somewhat inferior specific impulse. In the search for a fuel for the Titan, hydrocarbons were considered, along with various derivatives of hydrazine, a compound containing two atoms of nitrogen and four atoms of hydrogen (N_2H_4). Among the candidate oxidizers were nitric acid (HNO_3) and nitrogen tetroxide (N_2O_4). As a result of this search, nitrogen tetroxide was selected as the oxidizer. Two hydrazine derivatives, monomethyl hydrazine and dimethyl hydrazine, were identified as the best choices as fuels. The methyl group in these latter fuels is composed of a carbon atom with three hydrogen atoms attached (CH_3).

Making nitrogen tetroxide on Mars requires the use of nitrogen from the atmosphere and oxygen from carbon dioxide electrolysis, both of which are readily available. Making hydrazine requires the use of nitrogen and hydrogen, and the methyl hydrazines require nitrogen, hydrogen, and carbon. Finding a source of hydrogen is therefore essential to making any hydrazine-based fuel.

The most abundant atmospheric source of hydrogen is water vapor, but the prevailing low temperatures dry out the Martian atmosphere very effectively, keeping the water abundance usually below one part per million of the already thin atmosphere. To remove water vapor from the atmosphere, it is necessary to cool the gas below ambient temperatures to the point at which ice condenses directly from the gas, the so-called dew point (here it would be more accurate to call it the hoar-frost point). Since each liter of water vapor is dispersed through a million liters of carbon dioxide, nitrogen, and argon, the extraction of water vapor by refrigeration involves cooling vast volumes of atmosphere, at great expense. The magnitude of the task becomes clearer when we calculate that we can extract only 0.01 grams of ice from that million liters of atmosphere. Thus, to make ten tons of water available for processing use, we would have to remove all the water vapor from an astounding 10^{15} liters, which is 1,000 cubic kilometers of gas! This is a feat of mythic proportions, comparable to extracting gold from sea water—or helium-3 from the lunar regolith. Extracting water from the Martian atmosphere economically appears to be impossible. Where, then, can we find the hydrogen we need?

Since most of Mars is covered by permafrost, and since clay minerals and hydrated salts are probably nearly omnipresent, a concentrated source of water would be available to missions that land almost anywhere on the planet. During local spring in the polar regions, polar ice and perpetual sunlight may both be available simultaneously! The key to manufacture of storable fuels on Mars lies in the use of water-bearing surface materials, not atmospheric water vapor.

The time has come to recognize that reliance on the atmosphere as the sole source of Martian propellants gets us one foot firmly in the door but leaves the rest of us outside in the cold. It gives us the minimum performance we need to get off Mars but leaves much room for growth and improvement. Further progress beyond carbon monoxide–oxygen rockets requires the use of hydrogen and ultimately motivates the extraction of water from the planetary surface. This in turn requires new equipment of various sorts to be transported to the surface of Mars. First, we must excavate clay minerals, ice, or a mixture of the two (permafrost). Very cold ice is very hard, and very cold permafrost is extraordinarily tough to cut through. Second, we must heat that material to a high enough temperature to liberate water. This is easiest if we start by mining rather pure polar ice, and it is most difficult if we must extract water from clays or hydrated salts. Finally, we must include a new apparatus for integration of water into the manufacture of propellants.

An ingenious path to early integration of hydrogen into Martian propellant manufacture has been suggested by Bob Zubrin of Lockheed-Martin. He proposes initially bringing a small tank of hydrogen from Earth for use in making storable hydrogen-containing propellants on Mars. This removes the need for the relatively massive and failure-prone mining and extraction equipment needed to get water. But it also incurs the expense, weight, power demand, and complexity of the tankage, insulation, and refrigeration equipment needed to maintain the tank of liquid hydrogen during the nine month journey from Earth and deliver it safely to Mars's surface.

If we had a tank of terrestrial hydrogen available on Mars, what would we choose to do with it? Making hydrazine and its derivatives requires an additional piece of apparatus for capturing nitrogen gas from the atmosphere and a reactor to convert nitrogen into a more reactive product such as ammonia or nitric acid. Then another reactor would be needed to, for example, convert ammonia into hydrazine.

Alternatively, we might choose the simple path of introducing hydrogen into the chemistry of carbon dioxide, which we already have available. This can be done by means of a simple, well-tested device called a Sabatier reactor. Zubrin proposes using the Sabatier process to react hydrogen with carbon dioxide to make water vapor and methane. The water vapor can then be cycled by reacting it with carbon monoxide to make carbon dioxide and hydrogen. Thus, the hydrogen eventually all ends up as methane, the gas CH_4. Methane is a high-performance hydrocarbon fuel, better than kerosene or gasoline because of its very high hydrogen content, high combustion energy, and water-rich, low-molecular-weight exhaust. Methane, although a cryogen, is relatively easy to liquefy and store. Zubrin proposes burning methane with liquid oxygen for the return trip to Earth. The additional mass and complexity of the processing equipment to make methane (instead of stopping with carbon monoxide) and the penalty of having to carry the hydrogen tank from Earth offset the rocket-engine performance advantage. Since the carbon monoxide/oxygen rocket has adequate performance to carry out the return mission and is much simpler, its lower specific impulse is not a matter of critical concern.

If we elect to make hydrogen from water extracted from the Martian surface, then the need for terrestrial hydrogen and its insulation and refrigeration is removed. However, the mass of equipment required on Mars to extract and process water may be greatly increased. It is therefore important to find a Martian water source that requires the least mass of equipment and the minimum power consumption. After weighing all the alternatives, we may conclude that the easiest way to get the performance advantage of hydrogen-containing propellants is by using polar ice, not mining dirt.

It is interesting that the Sabatier process, although originally motivated by a desire to make moderate-performance but storable propellants on Mars, ends up in this version making a higher-performance, cryogenic propellant pair. This need not be so. For example, modest changes in the Sabatier reactor could permit it to make methyl alcohol (methanol, or CH_3OH) instead of methane. As every driver knows or suspects, methanol, as an oxygenated fuel, is inferior in its performance to hydrocarbons such as methane, but it is a liquid at normal Martian surface temperatures and requires no refrigeration for liquefaction or storage. This removes half of the cryogenics problem but leaves us with liquid oxygen.

Given such a system that uses carbon dioxide and water as its raw materials, we can consider any oxidizer (or fuel) that is made from only hydrogen, carbon, and oxygen. Aside from liquid oxygen, there is really only one other oxidizer worthy of consideration in this group:

hydrogen peroxide, a compound of two hydrogen atoms with two oxygen atoms (HOOH)—like water with an extra oxygen atom inserted. In a dilute solution in water, it is familiar as a household antiseptic. It is greatly inferior to liquid oxygen as an oxidizer, but it is a liquid over about the same range of temperatures as water (which means that, on Mars, it would freeze solid). But with a little insulation and a little heat, hydrogen peroxide can be maintained as a liquid on the surface of Mars indefinitely. The methanol/hydrogen peroxide combination has lower performance than the carbon monoxide/oxygen one, but the former is completely storable. Unfortunately, the performance of methanol/HOOH is so low that it is not satisfactory as a propellant combination for missions to return from the surface of Mars to Earth.

If, however, we also master the capture and use of nitrogen, then we can make nitrogen tetroxide, which not only is storable but performs reasonably well. Thus, the logic of stepwise development of propellant manufacture on Mars seems to lead us clearly to carbon monoxide/oxygen as the first step, with methane-oxygen or methanol-oxygen from terrestrial hydrogen as the second step, Mars hydrogen (water) use as the third step, and Mars nitrogen use to make nitrogen tetroxide as the fourth step. This sequence gives us considerable freedom in selecting possible products, including not only propellants but life-support materials and other commodities as well.

A variety of simple terrestrial life-forms can thrive in a medium of liquid water with carbon dioxide and ammonia available. We have already discussed the manufacture of hydrazine or ammonia from Martian nitrogen and water. Phosphorus is common in the Martian surface and presents no problems. Sulfates are abundant, and other sulfur compounds can be made as needed. These simple organisms, if protected from the wild daily extremes of temperatures experienced by unprotected sites on the surface, can serve as the bottom of a complex, multitiered food chain. Bacteria, algae, lichens, and molds are rarely obvious to us in nature, but their health is the basis for the strength of the overlying ecosystem. To the best of our knowledge, all of the needs of these simple but essential life-forms can be provided within a closed environment on the Martian surface. Photosynthetic organisms, beginning with creatures as humble as bacteria and blue-green algae, produce not only proteins and DNA but also oxygen for use by animals and higher plants. Further, many of these organisms destroy (indeed, feast on) materials such as hydrogen sulfide, hydrogen cyanide, and formaldehyde, which are toxic to higher life-forms.

Simply injecting familiar terrestrial life-forms into the surface environment of Mars will not work. Some freeze-dried organisms may remain inert but potentially viable for long periods of time, but the absence of liquid water and the presence of killing solar ultraviolet radiation and atomic oxygen preclude successful reproduction. Martian biological productivity and eventual self-sufficiency require domes or other enclosures to keep in warmth and keep out the deadly portion of the solar spectrum. Such environments, rudimentary closed ecological life-support systems (CELSS), can be charged with a supply of simple compounds of hydrogen, carbon, nitrogen, and oxygen, perhaps directly from the Martian surface, and then seeded with more advanced life-forms as soon as the dome's interior conditions permit. Experiments on Earth with such closed systems can and should be conducted. Interior temperatures and pressures inside insulated closed domes can easily be elevated relative to the Martian norm to provide conditions that are much more stable than on the exposed planetary surface. Such warmer, higher-pressure environments would be conducive to the introduction of hardy terrestrial life-forms.

This process of constructing habitable environments on Mars can be begun at once. The feasibility of changing the entire planetary climate to make it more Earth-like "outdoors" is open to serious question. The very low abundances of volatiles and especially the low nitrogen abundance all pose special problems. Adequate nitrogen is not only required by plants, but it also plays an extremely valuable role as a fire suppressant in habitations.

The entire planetary surface might be artificially warmed by introducing gases such as chlorofluorocarbons (CFCs), which are potent in enhancing the greenhouse effect. By this subterfuge, it should be possible to break the logic of the water–carbon dioxide greenhouse effect, which explains that "Mars is cold because it is cold." Added CFCs in trace amounts could raise the temperature enough to melt ice and raise the water content of the atmosphere by many factors of ten, reinstituting the natural greenhouse that apparently once maintained clement conditions on Mars several billion years ago, when the rivers ran and snowcapped peaks were mirrored in the seas.

The expense of making these CFCs on Earth and transporting them to Mars is prohibitive; however, the British atmospheric chemist James Lovelock has pointed out that the obvious solution to the problem is to make the CFCs on Mars out of indigenous Martian materials. Chlorides are present in the salts left by the vanished seas and by weathering of the crust. Fluorine is present in small amounts, as on Earth. The existence of fluoride ores is, however, uncertain.

The traditional term for making the surface of another planet more Earth-like, *terraforming,* was coined by the science fiction author Jack Williamson. The term evokes images of colonialist arrogance, a perversion of the state of a planet into something utterly alien to its nature—in William Burroughs's language, a "clockwork orange." But what we envision for Mars is an interference more akin to refurbishing and rewinding a fine old clock we have found that ran down many years ago. We are restoring Mars to where it was when it slipped off the rails of planetary evolution. Rather than terraforming, we are "areo-reforming" Mars—restoring Ares, the Greek god of war, to his rightful dominion. We cannot, and arguably should not, attempt to make Mars into another Earth. But we can give it new life, life that will prosper and proliferate in proportion to how well it is adapted to the restored, reinvigorated Mars. This principle applies to "human" life as well as bacteria; it foreshadows a *Homo sapiens martiensis.*

11

PHOBOS
AND DEIMOS

Everybody knew Kai Peterson was a little odd. I guess I've talked to him more than most, so I know his repertoire pretty well. You could say I've had my shots, and I'm no longer allergic to him. He's really a pretty decent sort when he isn't riding one of his hobbyhorses, but that's not very often. Maybe it was his thing about cultured seaweed that got to some people. Others caught a clue from his obsession with second-millennium mining songs. The rest based their opinions on his endless paeans to the future of Phobos.

Now, nobody else on Mars saw Phobos the way Kai did. Martians talk about Phobos very much the same way that Americans talk about Podunk, Australians talk about Tasmania, Canadians talk about Newfoundland, and Samuel Johnson talked about Scotland. Everybody knew—in fact, everybody learned from their Mentors—that Phobos and Deimos were a total bust. Two words: no water. No water meant no propellants. No propellants meant no cheap access and no metals. No free metals meant nothing to mine. Everybody—that is, everybody except Kai—has known since about 1990 that there was no water on Phobos. That was back in the Mad Millennium, with all its wars and crusades and genocides. Technically, the killer was that there was no 3-micron absorption feature in the infrared reflection spectrum of Phobos. Savvy? No 3-micron feature, no water, no ice, no hydroxyl silicates, no clays, no hydrated salts. Zip, nichtevo, nullo, blotto, nyet, zilch, niente, rien, nichts, and nada. Obvious to everybody but Kai.

So when I ran into Kai outside the Big Dipper in Old Town last night, under their big neon sign that says "Open 24.5 hours per day," I considered having an urgent prior commitment elsewhere. But it turned out Kai was in a good mood and buying, so I stayed.

"Yeah, I've some good stuff going," he admitted. "You know, out on Phobos."

I recoiled reflexively, but just then our celebration arrived. I resolved to give him ten full minutes of quality time and then excuse myself.

"Phobos, huh? Tell me all about it."

Kai leaped into the breach. "I've been saving up for fifteen years to get the propellant for a trip out there and back. Olympus Mines offered to let me use one of their oldest orbital freighters at cost, and I figured it real close: with no payload, one of their C-class orbital tankers could get to Phobos and back. So I cashed in all my savings—$300,000— and went." Kai munched energetically, organizing his thoughts. The poor slob wiped out his savings for a pipe dream. You gotta feel sorry for him. But he has a good job piloting with Olympus. Some day he'll recover. Let's see—seven minutes left on the mission clock, and the clock is running.

"So then what?" Prime the pump, and you'll be done pumping sooner.

"Well, I slept during coast phase of the outbound leg, woke up a couple of hours before encounter all refreshed, and did my apoapsis burn to match orbits with Phobos. I ran the usual suite of analytical runs as I coasted along. Covered the whole surface. Guess what I found?"

I tried to say "water," but the word stuck in my throat. Maybe it was the pretzels. I tried again, and words came out. "No water. Not one trace of water at all," I volunteered.

"Right!" Kai responded cheerfully.

I was left speechless for a few moments. Why was he so happy about the end of his silly dreams? "But—did you find something else that's really good?"

"Nope. It was a washout, just like in the textbooks. The whole surface was bone dry."

"So you came back . . . " I foundered. Four minutes left on my chrono.

"No, I landed. I put down as soft as a snowflake right between a pair

of crater chains. And I found that two of the bigger craters had holes in the bottom."

Now, those crater chains are funny. They're not made by impacts: they don't have raised rims. They look just like a row of funnels, ten, twenty, even thirty in a row in a nice straight line.

"You mean, like loose stuff running down into a buried crack?"

"Yep. Just like that. So I went down one."

I almost choked. "Alone? You went out of the ship alone, with no one on board to back you up, and you crawled into a cave on Phobos alone? What about all the safety rules? Are you totally crazy?" I didn't really regret the insulting words. They seemed justified. I even forgot to check my chrono.

"I didn't do any crawling. I used a kevlar safety line and my suit jets." He paused for a few moments, gazing earnestly, even plaintively, at my face. "Aren't you gonna ask me what I found?"

"Well, sure . . . "

"Holes."

"Holes?"

"Yeah. Giant cracks and fissures. About a quarter of Phobos is empty space. It's a giant rubble pile. You can tour the whole inside of it if you trail a safety line so you can find your way out again. And everywhere except in the surface dirt there's water. That surface dirt is dry because it it's been shock-heated over and over by impacts, and it's lost all its water. But the inside's another story. Clays. Hydrated salts. Twenty-one percent water! I filed my claim as soon as I got back. My estimate is $10,000 trillion."

I thought I would have a heart attack right on the spot.

"So I'm a rich man, Jerry, and all I need is a little cash to get started on the mining. Now, I like you, Jerry. You always gave me the time of day. No wisecracks and putdowns. So I was wondering if you'd like to be my partner . . . "

—Insider Trading

∴ ✱

Ares, the Greek god of war, rode into battle in a chariot drawn by two horses named Phobos ("fear") and Deimos ("dread"). When the

American astronomer Asaph Hall discovered the two moons of Mars in 1877, it was natural that he should draw on his classical education and name them after those mythological steeds.

Phobos and Deimos are small, irregularly shaped bodies, generally similar to carbonaceous asteroids in color, reflectivity, and density. Their masses are both on the order of a millionth of the mass of Earth's Moon. Their very low escape velocities, roughly ten meters per second, vary markedly from place to place on their highly nonspherical surfaces. Their surface gravity is so weak that astronauts on their surfaces could almost jump out of their gravity wells into independent orbits about Mars.

Both satellites pursue nearly circular, equatorial orbits about Mars. Phobos orbits at a distance of 9,378 kilometers from the center of Mars, and Deimos orbits at 23,459 kilometers. Their orbital periods (the "Phobos month" and the "Deimos month") are very short—only 0.40 and 1.23 Martian days, respectively. Thus, as seen from the surface of Mars, the moons cross the sky in opposite directions. Phobos is large enough and close enough to eclipse the Sun over a tiny area on the Martian surface as it passes through the daytime sky.

Mars synchronous orbit, MSO, is the altitude at which a satellite of Mars orbits the planet in one Mars day. That locates MSO between the orbits of Phobos and Deimos. Any satellite closer to Mars than MSO orbits faster than a Martian day, so any tidal bulge raised in Mars's crust by that satellite will tend to lag behind the satellite and gently retard its orbital motion through the gravitational attraction of the tidal bulge on the satellite. The loss of orbital energy by the satellite causes it to evolve slowly inward into a closer, shorter-period orbit. By Newton's laws of motion, this force also accelerates the rotational speed of Mars. Thus, the orbit of Phobos is constantly evolving closer to Mars and farther from MSO. By the same reasoning, the orbit of Deimos, with a period longer than a Mars day, is constantly evolving outward from both Mars and MSO.

Both Phobos and Deimos always keep the same side toward Mars. They are rotationally "locked on" to Mars, with rotation periods equal to their orbital periods.

From the surface of Phobos, a velocity increment of only 560 meters per second would permit descent into the Martian atmosphere, either to land on the surface of Mars or to aerobrake into low orbit about Mars (LMO). From Phobos, to escape from the Mars system altogether requires only 890 meters per second. Injection into an orbit that takes it all the way back to Earth requires 2,880 meters per second. For Deimos, in its higher orbit, only 560 meters per second suffices to escape from the Mars system. Return to Earth from Deimos requires a delta V of 2,550 meters per second.

Both of these tiny moons were studied by the *Mariner 9* Mars orbiter and by the Viking orbiters. They are both heavily cratered and covered by a deep regolith. Long chains of tiny craters on Phobos suggest the drainage of regolith into deep cracks that riddle its interior. Further, the largest craters seen on both satellites are nearly large enough to blast them into pieces. Debris ejected from either satellite by an impact goes into a belt of rocks and dust that closely follows the orbit of the target satellite about Mars. In effect, the debris is captured and contained locally by the Martian gravity field. The ultimate fate of most of this collision debris is to collide again with the satellite from which it was originally ejected. These collisions are very gentle, typically at speeds of only tens of meters per second, resulting in capture of the debris rather than further impact erosion.

Such reaccretion of collisional debris from Phobos and Deimos provides a deep regolith of satellite material that has been repeatedly shock-heated by impacts. Asteroids, on the other hand, spread their collisional debris over the incomparably larger space of their orbit about the Sun. Thus, only a tiny fraction of the debris eroded from asteroids by impacts is successfully reaccreted. Efficient regolith accumulation is therefore not possible on asteroids similar in size to Phobos and Deimos. We must expect that asteroids in this size range would have suffered serious erosion of their regolith.

Regolith formation is also made easier if the crushing strength of the surface rocks is low. Spectroscopic studies of Phobos and Deimos confirm a very low albedo (reflectivity) and a reflection spectrum similar to carbonaceous chondrites, the weakest class of recovered meteorites. The close flybys of both moons by the *Mariner 9* and Viking orbiters provided accurate photographic determinations of the shapes, diameters, and volumes of both bodies. Combined with estimates of their masses from the gravitational effects of these satellites on the passing spacecraft, this size information makes it possible to calculate their density. The answer is an astonishing 1.9 grams per cubic centimeter, even less than the 2.4 g/cm³ of the CI chondrites (see chapter 5). The implication seems to be that these moons are similar to CI material, but with 15 percent or 20 percent of internal voids due to a pervading system of wide, deep fractures. If the solids are drier than CI meteorites, then they must be denser, and the percentage of void space must be even higher.

All these factors suggest the importance of a direct search for water in the spectrum of these satellites. Water—whether present as solid ice, hydrated minerals, or hydroxyl silicates—contributes distinctive absorption features in the infrared part of the reflection spectrum. All three have different-shaped absorption features at wavelengths close to three micrometers. It was through the study of this portion of the spectrum that University of Arizona astronomer Larry Lebofsky was first able to demonstrate the presence of water on C-type and related dark asteroids. But the results of a study of Deimos by Lebofsky, Jeff Bell, and J. R. Piscitelli of the University of Hawaii are puzzling: no trace of water can be found in their spectra of Deimos. What could these bodies possibly be made of that is so dark and has such a low density, yet contains no detectable water?

One explanation, though still in the realm of informed conjecture, seems satisfactory: that the original material of Phobos and Deimos really was CI chondrites, from which most of the water has been baked out by repeated impact-shock heating. The dark color would be attributed to a mixture of partially pyrolyzed (charred) organic matter with magnetite, which was not hot for a long enough time to permit these materials to react with each other. Such a mixture might easily contain enough organic hydrogen to be equivalent to more than 1 percent water, as well as several percent extractable carbon and a fraction of a percent of nitrogen.

The principal alternative to the carbonaceous-chondrite hypothesis is that Phobos and Deimos are made of materials similar to ordinary chondrites, essentially devoid of water, which has been darkened by repeated shock-wave heating until it is a dark as carbonaceous chondrites. Indeed, according to University of Arizona meteoriticist Dan Britt, a number of heavily shocked "black chondrites" are known that combine the albedo of a C chondrite with the density of an ordinary chondrite (about 3.8). This scenario is plausible enough, but it requires that the Martian moons, which presently have densities close to 1.9, must be half empty space. A random assemblage of uncompacted angular fragments of a wide range of sizes, held loosely together by their mutual gravitation, would probably not be so empty.

It was first pointed out by Arthur C. Clarke in 1939 in the *Journal of the British Interplanetary Society* and developed in detail by aerospace writers Dandridge M. Cole and Donald W. Cox in 1964 in their book *Islands in Space* that the Martian moons were potential waystations and refueling depots on the route to the stars. If Phobos and Deimos are in fact similar to the carbonaceous chondrite meteorites, they would be very attractive refueling stops in the Mars system, providing propellant for landings on Mars and for return from the Mars system to Earth. A mission arriving in the Mars system could fire its rocket engines briefly just above the Martian atmosphere and be captured into an eccentric orbit about Mars that takes it out to Phobos and beyond. The mission could then use a further small amount of propellant to land on Phobos (or Deimos). A solar- or nuclear-powered processing plant that extracts water from the surface material could then make propellants and load enough into the vehicle's tanks to land on Mars and possibly even take off again.

If Phobos becomes a standard refueling base for operations in the Mars system, arriving vehicles would arrive with nearly empty fuel tanks and take on hundreds of tons of water for use in their solar-thermal or nuclear-thermal engines. The trillion tons of water suspected to be present in Phobos would suffice for several such refuelings per day for several million years.

Phobos and Deimos are promising way stations on the route between Mars and Earth. Their utility in servicing traffic from Earth to the asteroid belt is, however, dubious. A vehicle launched from Earth on an optimum trajectory to a belt asteroid would pass Mars at a sharp angle to Mars's orbital motion. The penalty incurred in matching velocities with Phobos, refueling there, and then getting back up to speed on the way to the belt would be severe. A better plan would be to launch from Earth on a trajectory optimized for transfer to Mars. Then the penalty paid in orbit matching with Mars and Phobos would be much smaller. But still it would appear to make little economic sense for an Earth-to-Mars vehicle to waste the fuel needed to stop in at Mars. Most near-Earth asteroids follow trajectories that are much better suited to the needs of belt-bound Earthlings.

There is another interesting use to which Phobos and Deimos could be put. Both are reasonably close to MSO, and their orbits in fact bracket MSO. Suppose that metallic iron alloy (such as piano wire), made as a byproduct of extraction of volatile elements from Phobos and Deimos, is collected at a platform orbiting in MSO. If we unreel a strand of this wire inward from our MSO station toward the Martian surface, the lower end of the wire will experience an increased gravitational attraction toward Mars. Since the wire is firmly attached to a much more massive object in MSO, it will orbit Mars at exactly the same rate as that of the satellite in MSO, but closer to the planet. The lower end of the wire cannot be in an independent orbit about Mars because its orbital period is fixed at one Mars day by the fact that it is anchored to the MSO platform: this means the tip of the wire not only experiences a larger planetward force than a freely moving satellite at the same altitude, but it also experiences a weaker centrifugal force pulling it outward from Mars. The net result is clear: the wire will be pulled toward Mars and will stay taut under this combination of forces.

A wire unreeled outward from the MSO platform, directly away from Mars, will, by the same reasoning, experience both an increased centrifugal force pulling it outward from Mars and a lesser gravitational force pulling it inward. Thus, a wire unreeled outward will also be pulled taut and will point radially outward from Mars.

Both the inward-directed and outward-directed wires exert a force on the MSO platform. Let us sidestep the complex problem of unbalanced forces by simply unreeling both wires at the same time at such rates that the *net* force on the MSO platform is zero. Now, how far can we unreel these wires? In the Mars system, there is no reason why we could not lower the inner end all the way to the surface of Mars. The weight of the wire stretching all those thousands of kilometers is of course great, but by proper choice of very strong wire materials and by appropriate shaping of the thickness of the wire (to make it thickest where the strain on it is greatest) we can in fact run a wire from MSO down to the surface of Mars. Since the MSO platform orbits perpetually over a fixed point on the equator of Mars, the end of the wire will rest at a well-defined fixed point on the Martian equator. Meanwhile, the upward-directed wire will stretch many thousands of kilometers above MSO. It must, of course, orbit once per Mars day because it is physically attached to the MSO platform. Its tip travels several times as far as the MSO platform in each revolution about the planet: its speed is far in excess of orbital velocity at its altitude. In fact, it is traveling well above escape velocity.

This remarkable device that connects the surface of a planet to escape velocity, appropriately called a skyhook, was first described in the scientific literature by Konstantin Tsiolkovskii in 1895 and was first brought to modern public attention in 1960 through an article by Yuri Artsutanov, a Soviet scientist, in, of all unlikely places, the Communist youth newspaper *Komsomolskaya Pravda*. No skyhook has ever been built: Earth's gravity is so strong that it requires almost impossibly strong construction materials, akin to flawless diamond fibers a hundred thousand kilometers long. The Moon rotates so slowly that synchronous orbit is not achievable. Of course, skyhooks on rapidly rotating, low-gravity asteroids would be easy to build and install.

Imagine a solar-powered mechanical ant with a tiny electric motor that starts out on the surface of Mars and begins to climb the skyhook. As it crawls up the cable, its weight steadily diminishes (less gravity and more centrifugal force) until it achieves weightlessness upon arrival at the MSO platform. Continuing to climb, the ant experiences a net outward force with every step. It is literally accelerated outward by riding the end of a very high speed whip. Now suppose that two fine metal wires run along the sides of the skyhook cable. Ants *above* MSO are running downhill and can use their electric motors as generators to pour power into the wires. Now ants on the lower leg of the cable, from Mars to MSO, can tap the electric power they need from the wires, and they no longer need to carry solar cells. An endless stream of self-powered ants can be sent up to the end of the cable with no further power input whatsoever!

What happens when the ants reach the upper end of the skyhook? They cannot simply accumulate there: eventually the strain on the skyhook would become so great that it would tear the cable apart. The alternative is simple: they let go. Depending on what direction they are heading in at the moment they let go, they will depart from Mars at a speed far in excess of the local escape velocity and streak out toward distant parts of the solar system. Ants that let go above the noontime point on the Martian surface will have a high velocity in the direction opposite to, and largely canceling, Mars's orbital motion. They will be traveling far too slowly around the Sun to stay in Mars's orbit, but they will instead drop inward to intersect the orbits of Earth, Venus, or even Mercury. Ants that let go from the tip of the skyhook at local midnight will have their speed added to Mars's orbital velocity and will streak outward from the Sun in elliptical orbits that graze Mars at perihelion but reach to the asteroid belt or the giant planets at aphelion.

Two obstacles block the way of physical realization of the Mars skyhook: Phobos and Deimos. On their present orbits, they would inevitably collide with the skyhook and damage or destroy it. Of course, if Phobos and Deimos were moved into MSO, they would no longer pose a threat to the cable. Instead, they would serve as an excellent MSO platform, full of raw materials for skyhook construction. There is no shortage of building materials: the two trillion tons of iron and nickel in the Martian moons would suffice to make a cable with a cross-section area of one square meter that would reach from Mars to the Sun!

Where would the energy come from to power this astonishing machine? The answer is simple: it taps the rotational energy of Mars. Hurling a millionth of the mass of Mars (six hundred trillion tons) from the tip of a skyhook with a length twenty times the radius of Mars would slow the equatorial rotation speed of Mars, now 22,000 centimeters per second, by a minuscule 0.4 centimeters per second. (This would lengthen the Mars day by 1.8 seconds.) For a skyhook of this length, the speed of the tip is 4.7 kilometers per second. Departing payloads could be flung as far out as the outer edge of the main asteroid belt (3.3 astronomical units).

In the long run, access to distant parts of the solar system should be very easy from Mars—certainly far easier than from Earth.

12

THE ASTEROID BELT: TREASURE BEYOND COUNT

To Mrs. Johnson's fourth-grade class at Roosevelt Elementary School, Springfield, Illinois:

Hi, kids! It's been over a week since I last wrote, so I thought it was time to bring you up to date and answer some of your questions.

Last time I wrote I was 18 weeks out from Marsport with my load of lithophage factories, and I had just made visual contact with my transit station, 2037 KD12. Now that's not a very exciting handle for an asteroid, but the Kid (as I call it) isn't very exciting to look at. Unless you know your asteroids, that is.

The Kid is a C-type asteroid in an eccentric orbit that brings it all the way in almost to Mars at perihelion, and all the way to the outer edge of the asteroid belt at aphelion. It covers the whole width of the belt twice in each of its 3.7-year trips around the Sun. There's been talk about hollowing it (and a few other carefully chosen asteroids) out and making it into a kind of traveling hotel, but that's still a way off. For one thing, nobody can decide who should pay for it!

Anyhow, the Kid is about three klicks wide and looks sort of like two lumpy potatoes stuck together end to end. So it's nearly four times as long as it is wide. Because it's a C-type, it's pitch black, really tough to see at a distance. In the infrared it glows like a ruby, but that's another story. Maybe Mrs. Johnson can tell you about that.

The crazy gravity field and the fast four-hour spin period persuaded me to set down on the summer pole, where several other visitors have landed in the past century. The Sun's faint out here, but at least at the summer pole it shines constantly, for about two Earth years at a time.

I wish you could have seen it as I was coming in on the leading side, near the dark south pole. I'll send along a couple of images from my file, but video is too expensive for a poor prospector! Anyhow, I drifted slowly past the Kid from pole to pole as it rotated beneath me so that I could do a complete chemical and physical map of the surface. Sure, the basic info's on file, but most of the good stuff is proprietary and owned by other prospectors who would never let it get out. This was my first time here; I've got great modern equipment; so I did the whole job. Within eight or nine hours I had completed the whole mapping run. By the way, you know it was dark around the winter pole, just like on Earth. So how could I map the part that I couldn't see? Simple—I used my radar! After I completed the mapping run, I lined up with the north polar axis and killed my orbital velocity to begin a vertical descent. A big burp from the engine and I was down on the surface.

The Kid is certified juicy. It has about 20 percent water, mostly in clays and salts, and about 20% other easily extracted volatiles, including organic gunk and oxygen extracted from iron oxides. You know what they're good for—life support, propellant, the works. And because the Kid has been visited before, there are already a bunch of lithophages at work. One of my buddies had dropped off a lithophage factory here five years ago.

It's on my flight plan to come here and pick up water, of course. If you don't have permission, that's piracy—and the penalty for piracy is death, not necessarily with the benefit of a trial. I signed up for 120 tons of water to refill my propellant tanks. According to my gauges, I really need only 98 tons, but it's better to plan in a safety margin. You don't want to be caught short! So I logged in and exchanged authorization codes with the local litho factory, then asked it to call in a bunch of lithos that had full tanks.

The local litho factory is still turning out new lithos after sixty-two months of operation. That means there must have been some 250 lithos produced already. Each one eats about three grams of rock per minute and makes about a liter of water per Earth day.

I checked out the litho factory that is part of my contract and found it in good shape. As another part of my deal with my buddy who put it here, I stuck a magazine of 500 brand new lithophile controller chips into the factory. Every other part of the new lithos is made by the factory out of local asteroid dirt. I ran the diagnostic software on the factory and changed out a few mechanical parts that were showing signs of wear. Factories are designed to be fault-tolerant, redundant, self-diagnosing, and very easy to repair. With my class-A license, I'm bonded to do repair work on the side. It's good money, even though the jobs are far apart!

People used to talk about making self-replicating machines—that means machines that can build perfect copies of themselves. That's illegal now, of course. You probably have learned about the Kansas City Panic in school. Now the factories and the lithos are designed so they can't make copies of themselves and can be shut down by remote control three different ways, and even shut themselves down automatically if they fail a radio check-in test.

Each litho carries only about 500 liters of water, so it takes a while to load 98 tons. I'll be done in about forty hours, and then I'm off to a long series of new asteroids. About every fourth asteroid will be a gas station where I can collect a tankful of water and do contract repair and maintenance work. On every asteroid that isn't already claimed, I'll put down one or two litho factories and check them out before moving on. Over the next four years I will visit ten new asteroids and three that already have lithos at work (one of which belongs to me). Each one we catch becomes a jumping-off point for dozens more.

I have a couple of questions here that you sent me. Benny asked about whether there wasn't a problem about using up all the water in these asteroids by refueling from them. Basically, Benny, it takes about 100 tons of water to refuel. This asteroid contains enough water for about 100 million refuelings. I just checked the log in the litho factory, and this is only the third time this one has been used.

Sally asked why my messages are always so short. That's because I (and all my tagged asteroids) get a daily transponder check to see if we're where we said we'd be and to make sure we're all OK. At that time I can dump a short message file free of charge. Anything beyond that requires special arrangements and costs a real bundle.

It's lonely out here, and I'm sure glad to have someone to talk to! I'll write again soon, and I hope you'll do the same.

Your pen pal,
Ed "Fuzzy" Levecque
Belt Prospector Lic. A:L337-01

—Beltstrapping

Science fiction is replete with tales of irascible old asteroid prospectors and zero-gravity miners, neither noted for their social graces or refined standards of personal hygiene. The asteroid miners of E.E. "Doc" Smith, Poul Anderson, and other science fiction authors are solitary frontiersmen and rugged individualists. In fact, the fictional world of the asteroid belt seems hauntingly familiar, with its lonely, hazardous, and often concealed minesites, its decadent boom towns with grossly inflated prices, and its boisterous miners in town for a few days to pick up supplies and go on a bender. There is usually an elaborate entertainment industry incorporating slick gamblers, painted women, and a variety of dubious establishments pandering to different tastes and offering countless forms of dissipation. Looming in the background are always the twin prospects of sudden incalculable wealth and instant agonizing death.

But we can perhaps do better than merely adding rockets and blasters to old western movies: in many ways, we can already anticipate how and why people will first get to the asteroid belt, what they will search for and what they will find, and how they may make themselves self-supporting in their endeavors. This story is vastly more complex, diverse, and entertaining than its fictional counterparts.

The first foot in the door of the belt is afforded by missions to the near-Earth asteroids. As we saw in chapters 5 and 6, most near-Earth asteroids (NEAs) follow elliptical orbits that drop briefly inside Earth's orbit about the Sun at the time of perihelion passage. These asteroids then coast away from the Sun on long ellipses, slowing as they climb outward against the Sun's gravity. At their farthest point from the Sun, at aphelion, most NEAs find themselves deep in the heart of the belt. Any equipment working on these bodies (and any crew present to install or service that equipment) will get a free trip to the belt every few years as the asteroid pursues its orbit about the Sun. It is relatively easy to ride an NEA out from Earth to the belt, and once there to transfer to any of tens of thousands of large belt asteroids.

The ancient human desire to visit and settle Mars may also play an important role in giving us access to the belt. Several suggestions have been made regarding the use of an NEA in an Earth-grazing orbit as a "traveling hotel" to move perpetually back and forth between the orbits of Earth and Mars. Cycling bases that repeatedly pass by Earth and Mars are most useful if they have orbital periods that are close to resonant relationship with both planets. The simplest example is an orbit that has the same period as Mars and that always intersects the orbit of Mars outbound from Earth, every 1.88 years just as Mars is passing by. Such an orbit, if it just grazes Earth's orbit at perihelion, will reach out to 2.05 AU at aphelion. This is so close to the inner edge of the densely populated asteroid belt (2.2 AU from the Sun) that reaching the belt from the cycler is easy: many thousands of large asteroids are now within reach.

The narrow confines of the inner solar system seem claustrophobic compared to the asteroid belt. Little airless Mercury and the parched, blazing pressure-cooker of Venus lie only 0.4 and 0.7 AU from the Sun, well inside Earth's orbit. It is Earth's distance from the Sun that defines the yardstick of distances in the solar system: the mean distance of Earth (and its Moon) from the Sun is 1.0 AU. Mars follows its eccentric orbit at a mean distance of just over 1.5 AU from the Sun. These five large bodies are accompanied by a diffuse swarm of smaller bodies, the NEAs plus the two tiny moons of Mars. Comets that spend most of their lives far from the Sun do not loiter in this region but occasionally dash through in splendid haste, successfully avoiding collision about 99.9999999 percent of the time. (Perhaps, like Parisian motorists, they believe that it is hazardous to drive in densely populated areas and therefore, in obedience to the Cartesian dictates of pure reason, accelerate to high speeds so as to spend as little time as possible in the heavy traffic.) About two thousand kilometer-sized NEAs are orbiting in the terrestrial planet region, and nearly a thousand comets have passed through it in historical time.

The belt, however, consists of perhaps forty thousand bodies larger than a kilometer in diameter. The largest, Ceres, is about a thousand kilometers in diameter, implying a mass of about 2.5 percent of the mass of Earth's Moon. The total mass of all the asteroids, both known and undiscovered, must be close to 5 percent of the mass of the Moon. Most lie between 2.2 and 3.3 AU from the Sun, the region referred to as the asteroid belt. They therefore lie well beyond Mars, but well inside the orbit of Jupiter, which is 5.2 AU from the Sun.

The orbits of the belt asteroids are by no means uniformly distributed over the volume of space between Mars and Jupiter, or even over the interval from 2.2 to 3.3 AU where most of them reside. Instead, many are strongly bunched according to their *mean* distances from the Sun; that is, they form groups with closely similar orbital periods. But the average asteroid has a distinctly noncircular orbit and varies in its distance from the Sun by about 10 percent above and 10 percent below the mean on each trip around the Sun. This means that these "families" of asteroids would not be at all evident in a snapshot showing the positions of all the asteroids at any moment in time. Rather than forming distinct, narrow bands of particles, similar to the thin threadlike rings of Uranus, they form a thick, diffuse disk about the Sun. Furthermore, unlike some science fiction portrayals, these orbital families are most certainly not *clusters* of asteroids that travel around the Sun as a group. They in fact move independently, with different orbital speeds and periods. Pairs of belt asteroids with nearly identical orbits rarely get anywhere near each other and are frequently 5 or 6 AU apart, as far as the distance from the Sun to Jupiter. They are families, then, only in the sense that they have similar distances from the Sun and similar orbital periods.

Discrete orbital families of asteroids occur at several locations outside the main belt, such as the Cybeles at 3.4 AU, the Hildas at 4.0 AU, and the Hungarias at 1.9 AU—but these contain only a tiny percentage of the number of main-belt bodies. On Jupiter's orbit about the Sun, about sixty degrees ahead of Jupiter and sixty degrees behind, centered on the apices of equilateral triangles that have the Sun at one corner and Jupiter at another, are two diffuse "clouds" of asteroids called the Greeks and the Trojans, bearing the names of characters from the *Iliad*. These Homeric bodies may be more numerous and massive than the entire population of the main belt, but they are so black and so distant from the Sun that they are hard to find and, once found, hard to study. They are also much less accessible from Earth than the other families of asteroids. Members of these clouds are not even especially accessible to each other: each of the two clouds fills a volume of space comparable to the entire inner solar system. The two clouds are about 9 AU apart, as far from each other as Earth is from Saturn.

Within the main belt are several distinct orbital families of large asteroids, each family named for its most prominent member. At the inner edge of the belt, 2.2 AU from the Sun, is the Flora family. Slightly farther out, near 2.4 AU, is the Phocaea family. Both the Flora and Phocaea asteroids have orbits that are inclined only a few degrees relative to the solar system's central plane. Three more prominent families in the main belt are the Koronis family, at 2.85 AU from the Sun and with orbit inclinations of only 2 degrees and eccentricities of only about 0.05 (they vary from their mean distance from the Sun by only about 5 percent as they circle in their orbits). The Eos family, near 3.0 AU, have typical inclinations of about 10 degrees and eccentricities of 0.17. Finally, the Themis family, centered at 3.13 AU, have eccentricities of about 0.15 and inclinations of only 1 degree.

It is not completely clear what significance these family groupings have. It is possible, for example, that most of the asteroids in one of these families are literally related to one another through collisional destruction of a larger asteroid early in solar system history. In a sense the gaps in the belt are better understood than the families: the gaps in the distribution of asteroid orbits correspond to orbital periods that are harmonics of the orbital period of the asteroid belt's giant neighbor, Jupiter (periods of one-fourth, one-third, two-sevenths, and so on of Jupiter's "year"). Asteroids in these "resonant" orbits, like particles of sand on a vibrating drumhead, are systematically driven away from "resonances," those locations where the disturbing gravitational forces of Jupiter and Saturn are periodic, and hence build up from orbit to orbit, reinforcing each other. At other, nonresonant locations (like the nodes on a drumhead), the perturbations on successive orbits are more or less randomly directed and tend to average out over long periods of time. Thus, Jupiter's gravitational pumping of the belt serves to clean out well-defined gaps corresponding to orbital periods that are harmonically related to Jupiter's orbital period. As a result, the momentary distance of the asteroid from the Sun is of no consequence; the *mean* distance, which defines its orbital period, is all-important. The fate of belt asteroids that wander into Jupiter resonances is particularly interesting: they will suffer severe orbital perturbations that increase their orbital eccentricity, causing them to begin crossing the orbits of the terrestrial planets. This is one of the two principal sources of new NEOs. The other half of the NEOs are extinct comet nuclei.

A simple test of the possible "genetic" connections of asteroids within a given family is to see whether each family has a distinct composition. If they are all alike in composition, then it becomes very reasonable to attribute such a family to the collisional destruction of one ancient asteroid. It is perfectly reasonable to wonder whether such a collision should produce an asteroid family with *two* different distinctive compositions: after all, it takes two asteroids to have a collision! But any large asteroidal target is hit vastly more often by small projectiles than by equal-sized asteroids: a small projectile traveling at the relative speeds of two belt asteroids can shatter a collision partner 100 to 1,000 times its own mass and then contribute only 0.1 percent to 1.0 percent of the mass of the debris cloud. Evidence of the second, smaller collision partner would then be very hard to find.

Two very different ways of determining the belt's composition exist. The first and older way is the study of meteorites. Meteorites have been falling to Earth throughout historic time. Indeed, according to the evidence of impact craters on Earth, the Moon, Mercury, Venus, and Mars, they have been falling as small stones, house-sized boulders, and flying mountains for the entire history of the solar system. The larger bodies are of course much rarer: a kilometer-sized asteroid strikes Earth on the average about once every 150,000 years. But small rocks fall from the sky much more frequently.

What, then, do studies of meteorites tell us about the materials of the belt? The belt contains many distinct kinds of material, each arranged in a ring about the Sun with a preferred average distance. These bands of material overlap each other considerably, but not enough to obscure the obvious zoning of the belt into concentric bands. At the inner edge of the belt we find S-asteroid materials similar to those meteorites that formed at high temperatures, solidifying and crystallizing from the molten state. Many of these materials are what we would call igneous rocks if we were to find them on Earth. While these rocks were still liquid, the materials in them were separated by the gravity of their parent asteroid according to their density. The densest liquid, a solution containing molten iron and iron sulfides and chemically similar elements such as nickel and cobalt, settled out in the gravitational fields of hot asteroids to form large pockets or central cores of metal and sulfides. This is the source of the fragments that we know as iron meteorites. Separation of metals from the melted rock leaves behind a silicate rock that is mostly made of silicates of magnesium and iron, with important amounts of the familiar terrestrial mineral plagioclase, an aluminum silicate mineral containing sodium, calcium, and potassium.

We saw in chapter 5 that only five abundant materials exist in most meteorites: iron- and nickel-rich metal, an iron sulfide called troilite, two iron-magnesium silicates called pyroxene and olivine, and plagioclase feldspar. All of these except troilite are common on Earth: Earth's core is mainly composed of metal and sulfides; the mantle is dominated by the ferro-magnesian silicates; plagioclase feldspar is ubiquitous in the crust. In fact, the structure of Earth is a straightforward example of the separation of melted minerals according to density. This separation of minerals by density, called geochemical differentiation, seems to be a universal feature of the rocky planets and S- (stone) and M-type (metal) asteroids.

But most of the meteorites that fall on Earth have never been heated strongly enough to melt and differentiate. They are extremely ancient, showing no evidence of heating to the melting point, with ages close to 4.55 billion years. They are fine-grained mixtures of minerals that in many cases have not even been warm enough to come to equilibrium with their neighbors. Almost all of these ancient meteorites contain dropletlike spherules of partially (or extensively) recrystallized glass. As discussed in chapter 5, these are called *chondrules*, after the Greek word for "seed" or "droplet." The meteorites that contain chondrules, called *chondrites*, account for about 85 percent of all the meteorites that are observed to fall. The chondrites are virtually unprocessed planetary raw materials, the stuff that failed to join a planet during the early days of our solar system. These chondrites belong to three major groups that differ in their composition. Most common are the *ordinary chondrites*, which contain the same basic minerals but in slightly different proportions. The ordinary chondrites, spectrally similar to the Q-type asteroids, come in three varieties that have different proportions of silicates versus metals and sulfide. These are called the high-iron (H), low-iron (L), and very-low iron (LL) chondrites. The ordinary chondrites make up more than 80 percent of all the chondrites.

The second main group of chondrites contains essentially no oxidized iron. Metallic iron and iron sulfide are abundant, and the principal other minerals are a pure magnesium pyroxene called enstatite and plagioclase feldspar. These meteorites, the *enstatite (E) chondrites*, make up only a few percent of the meteorite fall rate on Earth. Their parent asteroid (or asteroids) remains unknown.

The third and final category of chondrites, also described earlier, is a stunningly complex mixture of fine-grained clays, magnetite, water-soluble salts, and organic matter called the *carbonaceous (C) chondrites*. These carbonaceous chondrites fall into several families that differ in their degree of oxidation and in the amount of water, carbon, and other volatile materials in them. The most extreme types, the Ivuna-type (CI) and Murchison-type (CM) carbonaceous chondrites, contain from 5 percent to 20 percent water, up to 6 percent organic matter, and 6 percent sulfur (in the form of sulfates, elemental sulfur, organic sulfur compounds, and sulfides). The organic matter is extremely old and quite dissimilar to biological material. It was almost certainly produced by energetic nonbiological processes such as ultraviolet irradiation or lightning during the very earliest days of the solar system, before the planets formed. Some of it even predates the formation of the solar system.

The total amount of easily extracted volatiles in them (gaseous compounds of hydrogen, carbon, oxygen, nitrogen, sulfur, and so on) runs as high as 40 percent of their total mass. The solid residue after these volatiles are baked out is about 30 percent iron-nickel metal. They seem almost too good to be true—a rich source of both volatiles and metals!

Incredibly, more than half of the belt seems to be made of the ingredients of carbonaceous chondrites, mixed in varying proportions in several different asteroid classes. Half of the asteroids in the belt, including virtually all those in the outer half of the belt, belong to the C (carbonaceous) class of asteroids. We need only think back on our visits to the Moon and NEAs to appreciate that water and the other volatiles provide us with rocket propellants, radiation shielding, drinking water, atmospheric gases to breathe, and nutrients for hydroponic or dirt farming. Thus, the C asteroids and their related classes (D, P, B, G, etc.) are broadly available to those who venture into the belt for business or pleasure.

With this appreciation of the nature of the belt, it is appropriate to dedicate the remainder of this chapter to describing how we can use the resources of the belt to liberate the greatest resource in the entire solar system. To begin this task, we must first crudely inventory the materials available in the belt. It would be tedious to carry out such an inventory for every chemical element. As an example of the process, we shall concentrate on a single element whose availability is essential for industrialized society: iron.

The wealth of materials available in the belt asteroids, revealed by the study of meteorites, staggers the imagination. There is enough raw material in the asteroids to support a truly vast population. Consider that the belt contains some 7.5×10^{23} grams or 1.65×10^{21} pounds (1,650,000,000,000,000,000,000 pounds, or 825 quintillion tons) of iron. So extraordinary is this fact that we shall approach it from several perspectives to try to bring its enormity within our grasp. We shall look at the magnitude of the belt supply of iron from six different perspectives for the sole purpose of helping us to imagine its immensity. These are *not* six different serious proposals for using the metal resources of the belt!

Perspective 1: To raise the standard of living of the people of Earth to present-day North American, Japanese, or Western European levels we need about two billion tons of iron and steel each year. With the asteroidal supplies of metal at hand, we could meet Earth's needs for the next four hundred million years—even if we stupidly ignore recycling altogether!

Perspective 2: Earth is "running out of resources." Suppose we were to extract all the iron in the belt and bring it back to Earth. Spreading this amount of iron uniformly over all the continents gives us a layer of iron 800 meters (half a mile) thick. Allowing 10 centimeters (4 inches) of iron for each floor of structure, this is enough iron to cover all the continents with a steel-frame building 8,000 stories (80,000 feet, or 15.2 miles) tall. The building would be so large that there is just enough air on Earth to fill it to normal pressure levels.

Perspective 3: Earth's population wants to expand into space. We have enough asteroidal iron to make a metal sphere 920 kilometers (550 miles) in diameter. Hollowed out into rooms with iron walls, like a gigantic city, it would make a spherical space structure over 2,000 kilometers (1,200 miles) in diameter. Allowing 300 cubic meters for each resident, a family of five would then have 1,500 cubic meters (54,000 cubic feet). With a nine-foot ceiling, we could provide each family with a floor area of 3,000 square feet for private residential use and still set aside 3,000 square feet of public space per family. This artificial world would contain enough room to accommodate more than 10^{16} people. And how can we grasp what 10^{16} people means? That's 10,000,000,000,000,000 people—ten *quadrillion* people. Very simply, that is a million times the ultimate population capacity of Earth; a *million Earths of resources and room.*

Perspective 4: Plans for terraforming Mars and making it habitable have run into some problems: to conserve water and warm the surface we must, as we saw in chapter 10, build a dome over the entire planet. The problem is that we need a huge amount of metal to build the structure, as much as a half gram of metal for each square centimeter of surface area of Mars—an incredible 7.3×10^{17} grams (7.3×10^{11} metric tons, or 0.7 trillion English tons). No problem; that is one *millionth* of the amount of metal available in the belt. Fixing Mars would not consume enough resources to interfere with any of our really ambitious plans.

Perspective 5: Venus, with four times the land area of Earth, is unfortunately as hot as a self-cleaning oven and drier than the Sahara. To cool it off and make it habitable, we need sun screens to drop the intensity of sunlight striking its atmosphere several-fold. We then need to import a large amount of water—say, 10^{24} grams (a quintillion tons). The total amount of water in the belt would just about do the trick, but transporting it to Venus wastes almost all of it as propellant and leaves us none to work with in the belt. A large comet would do the job, but even a twenty-kilometer body (one hits Venus about once per billion years) would provide only 0.0004 percent of the amount of water we need. The only way to get enough water is to find and divert several Chiron-sized giant comets by steering them through close Jupiter or Saturn swingbys. This is more than a little hazardous, since it requires crossing Earth's orbit on the way to Venus. If you could get permission for this modest proposal, all you would need to do is find the funding. Estimates of the cost of the project run around 10^{16}. The amount of metal needed is ten times what we used on Mars. But that is only a hundred-thousandth of the amount of metal in the belt!

Perspective 6: People make themselves at home throughout the solar system. Our supply of iron is divided up among ten billion space habitats, each carrying one million residents. Each crew picks a location where it can make a living and settles there.

But what is all this iron worth? Why don't we just do the easiest thing and calculate what this amount of iron and steel is worth at present Earth-surface prices? Let's see—that's about $50 a ton times 7×10^{17} tons. Hmmm. Looks like 3.5×10^{19}. Now what's that in everyday units? Well, a billion dollars is 10^9, and in recent years the U.S. deficit has been about $300 billion, or 3×10^{11}. Let's see, how about the U.S. national debt? That's about $3 trillion, which is 3×10^{12}. So it's 10,000,000 times as large as the national debt! Let's look at this number another way. Let's divide this number by the total number of people on Earth to get some idea of the wealth generated per capita by using asteroidal iron and steel: that comes to $3.5 \times 10^{19}/5 \times 10^9$, or *$7 billion per person.*

I do not want to leave the impression that enough mineral wealth exists in the asteroid belt to provide $7 billion for each person on Earth. That would not be fair. In fact, this estimate completely ignores the value of all the other ingredients of asteroids besides iron. We know, for example, that for every ton of iron in the asteroids, there's 140 pounds of nickel. That comes to about $6 billion worth of nickel. Meteorite metals contain about 0.5 percent cobalt, which sells for about $15 a pound. That gives another $26 billion each. The platinum-group metals, which sell for about $460 per troy ounce ($15 per gram, or $6,800 per pound) make up about fifteen parts per million of meteorite metal. That comes to another $1.6 × 10²⁰, which is $32 billion per person. So far that is about $72 billion each, and we are not close to done. Add in gold, silver, copper, manganese, titanium, the rare earths, uranium, and so on, and the total rises to over $100 billion for each person on Earth.

It appears that sharing the belt's wealth among five billion people leads to a shameless level of affluence. Each citizen, assuming he or she could be persuaded to work a forty-hour week, could spend every working hour for 70 years counting $100 bills at the rate of one per second (that's $360,000 per hour) and fail to finish counting this share of the take. If we were instead to be satisfied with an average per capita wealth comparable to that in the upper economic classes of the industrialized nations today, roughly $100,000 per person, then the resources of the belt would suffice to sustain a *million times* as many people as on Earth. These 10^{16} people could all live as well as a ninety-fifth percentile American of the late twentieth century. With recycling and an adequate source of power, this immense population is sustainable into the indefinite future. The best use of the wealth of the asteroid belt is not to generate insane levels of personal wealth for the charter members; the best use is to expand our supply of the most precious resource of all—human beings. People embody *intelligence*, by far the most precious resource in the universe and one in terribly short supply.

The major missing factor in this discussion has been energy. How could we possibly provide enough energy to maintain the more extreme populations just calculated? "Fossil fuels" are indeed present in the belt, since carbon makes up about 6 percent of the carbonaceous asteroids. The total mass of carbon works out to about ten tons per person. Even if oxygen were free, the energy needs of a modern person (about one kilowatt average) would exhaust all the available carbon in only thirteen years. Then we would have the problem of what to do with the thirty-seven tons of carbon dioxide we made by burning that carbon! If we had an energy source capable of recycling this waste gas into carbon and oxygen, then we would use that energy directly and not waste it on carbon dioxide recovery. Burning "fossil" fuels is a really bad idea.

What about nuclear reactors? Asteroids contain a variety of rare radioactive isotopes of potassium, uranium, thorium, rubidium, and so on. Of these, uranium and thorium can be used to generate power in nuclear reactors. To get the most out of the mined uranium and thorium, they must be cycled through breeder reactors. But that makes plutonium, the ideal material for small nuclear weapons. The amount of uranium in the belt is about four billion tons, enough to make roughly a trillion tactical nuclear weapons. It would be a terrorist's heaven. The total energy content of all the radioactive material in the belt is almost exactly the same as the amount of energy given off by the Sun in *one second*. The energy needs of 10^{16} people, using one kilowatt each, would exhaust the total energy content of the uranium and thorium supply in about two hundred days. This discussion ignores the problems of induced radioactivity in reactors and the enormous problem of disposing of billions of tons of high-level radioactive waste in a heavily populated solar system. It also ignores both accidental and intentional releases of radioactivity and diversion of breeder materials for weapons use. So fission power is an even worse idea.

But wait a minute! Why not use solar power? The Sun pumps out power at the prodigious rate of 4×10^{33} ergs per second, equivalent to 4×10^{26} watts. Our supercivilization needs 10^{19} watts to keep going. The Sun is pumping out forty million times as much power as we need! But what do we need to do to capture and use that energy? The simplest answer (not necessarily the best—there may be even more desirable options that we have not thought of yet) is to use vast arrays of solar cells to convert sunlight into electrical power. If the cells have an efficiency of about 20 percent, similar to the best commercial cells made at present, then each square meter of cell area exposed to the Sun near Earth's orbit would generate 270 watts of electrical power continuously. We would need thirty-seven billion square kilometers of solar cells to provide our power needs, an area comparable to the total surface area of our habitats. At about 0.1 grams per square centimeter for the solar cells, we would need about 3.7×10^{19} grams of silicon to make the cells and perhaps three times as much metal to provide the supports and wires for the power-collection system. The asteroids give us 4×10^{23} grams of silicon, more than ten thousand times the amount we need for this purpose. The cost of the solar power units is set by the need to construct a few square meters of solar cells per person. The cost would be about two hundred dollars per person at present prices, or a few dollars per person at future mass-production prices. That is not your monthly electric bill: it is a one-time-only expenditure to provide all the electric power you will need for the rest of your life.

All this reckons with 1997 technology. New types of high-efficiency solar cells made of gallium arsenide or other exotic materials, combined with ultra-lightweight parabolic reflectors to collect and concentrate sunlight onto small areas of these cells, promise to perform much better than these highly conservative estimates.

So what should we do with all these resources? Where should we use them? Bringing some home is fine, but bringing them *all* home to Earth is, as we have seen, out of the question—we have no place to put them! Transportation of asteroidal material throughout the solar system is conceivable, but not obviously desirable. A carbonaceous asteroid containing 20 percent water can be processed to make a mass of liquid hydrogen and liquid oxygen equal to the original mass of water. Burning that propellant combination will accelerate the remaining mass of the asteroid to a speed of only one kilometer per second. Thus, complete exhaustion of the water content of a water-rich asteroid is not sufficient to change the location of the asteroid in the solar system by a significant amount. Certainly, one could envision using a portion of the water content to move small masses freely about the solar system, but there is no prospect of using this approach to move any significant portion of the mass of the belt to other locations such as Earth. The belt provides vast material resources, vast amounts of solar power, and vast elbow room. It would make a fine place for a lot of people to live.

Of course, once the belt becomes crowded, many people will want to move on to the unspoiled, uncrowded suburbs. That is when we move on to the Trojan asteroids, out on Jupiter's orbit about the Sun. Recent estimates suggest that the total mass of material in the Trojan and Greek regions (sixty degrees ahead of and behind Jupiter on its orbit about the Sun) is about three times that in the Belt. The abundance of volatiles may be five to ten times as large, since all these distant asteroids appear to be supercarbonaceous. Operating near Jupiter, 5.2 AU from the Sun, means we need a twenty-seven-times-larger expanse of solar cells for each citizen. But that is still far less than 1 percent of the amount of silicon actually available.

To recapitulate: We have used the resources of the belt to decompress Earth to a few billion people, terraform Mars, develop the belt and the Trojan region, and grow an affluent society of several tens of quadrillions of people. But we have barely mentioned Venus, with that unthinkable 10^{16} expense for building it a sunshield and terraforming it! How could we ever raise the funds to pay for such an outrageously expensive endeavor? Spreading the cost over all of humanity would mean about thirty cents each. Why not have the children hold a bake sale?

We need not judge in detail exactly what kind of a future we want for our descendants. We cannot, even if we wanted to, specify the political, religious, and philosophical beliefs and practices of our progeny. The question before us is much simpler and more general: *What can we do to give them a future in which they have the options to choose their own way?* Or perhaps the question really is whether we *dare* liberate ourselves from our familiar, cozy, but ever-more-crowded prison. In more poetic terms, as Walt Whitman said:

Darest thou now O soul,
Walk out with me toward the unknown region
Where neither ground is nor any path to follow?

The answer to Whitman's question is clear. Tsiolkovskii pointed out the obvious nearly a century ago:

Earth is the Cradle of Mankind,
but one cannot stay in the cradle forever.

We *can* choose to break the surly bands that tie us to Earth—if we dare.

13

THE OUTER PLANETS:
POWER BEYOND MEASURE

Note: The following is an unedited transcript.

All right, let's get started. Welcome to the Science Plus/NASA press conference. We, uh . . . is this microphone on? OK, let's see. Introductions first. I'm Mitch Garneau, mission manager for the Uranus Floater project. This is Devendra Dai, who's NASA's chief scientist for the mission. The suit is Jay Hagelin, the CEO of Science Plus, and next to him is Dorothy Cheng of code SL at NASA—uh, that's Solar System Exploration, the folks who buy the data we bring back.

Well, you all know why we're here. Those of you with long memories will recall just a little over seven years ago when we blasted off from Uranus. What? OK, a little more detail. Fine.

Where to start? OK, look, the way NASA has been doing business since the turn of the century is basically that they don't do most science missions themselves. Instead, they contract out—they send out a request for bids on some project the NAS thinks—no, not NASA. NAS is the National Academy of Sciences. Anyhow, something the NAS tells NASA is important. NASA draws up a shopping list of the data they want from a particular mission, the measurements they need and so on, and we— whoever wins the bidding competition, comes in under the cost cap, and can technically do the job—gets the contract. Twenty years ago—yeah, twenty years ago—we, now called Science Plus, formed a new startup company to propose the Uranus Floater mission to NASA in response to one of their requests for proposals. We got some venture capital, some aerospace engineers, some research types from universities and from the NASA centers that were closed—Ames and JPL—and we put together an industry team with some top-notch partners like Lockheed-Martin, Ball, ARCO, General Mills, and Fluor Daniel.

So, anyhow, we proposed a mission to land an atmospheric floater, basically a big hot-air balloon, on Uranus and use it as a platform to study the composition and structure of its atmosphere and its weird weather. You all know Uranus sort of lies on its side and the poles are warmer than the equator. This screws up the weather something awful. Also the composition is important because the thing that tipped over Uranus's spin axis was a really big impact. What? How big? Probably something the size of Earth or a little larger, right, Dorothy? Yeah. Well, look, let's not get into that right now. Dorothy will be glad to answer questions about the science end of the mission later, OK? Anyhow, the chemical signature of the impactor might be easy . . . OK, later, later.

So we sent our floater payload out to Uranus. It took about seven years to get there. That was—arrival was eight years ago. We decided to bring back a sample of the atmosphere. Now, if you were around then, you know there was a pretty big flap about that. Sample return from Uranus isn't easy, and it wasn't part of NASA's original plan. Some loud—some critics said it made the mission too big and too expensive, but look, there was a cap on the mission cost in the contract, right? We came in the cheapest of the three teams competing, and we won fair and square. Of course, some people said we were crazy to bid $86 million—remember, the cap was $105 million—and then spend $58 million on the launch vehicle. But we said we'd do it, and we're here today because we did it.

What? Could you repeat the question? What was the cost of the payload? Yeah, that was about $250 million. Please, could you quiet down? Let's save the questions for the end of my presentation.

OK, OK, take it easy. Yes, we spent $308 million to win an $86 million contract. Aerospace is real competitive these days. Could we hold it down so I can continue? It'll all be clear by the time I'm done.

Anyhow, in a few minutes—Devendra says three minutes—yeah, three minutes from now our return capsule will be skimming the top of Earth's atmosphere and braking down enough to be captured by Earth's gravity. In that capsule are more than ten tons of liquefied gas from the *atmosphere of Uranus. I said, please hold it down. All right, all right. What is it? Why do we need ten tons, and isn't a few grams enough? Yes, a few grams—a few liters (a liter is about a quart, you neolithic idiot—sorry, sorry). Yes, we returned a few liters of pure Uranus atmosphere as a bonus for NASA. We'll be delivering that sample gratis to the NASA program monitor on Space Station Delta in a few days. Yes, gratis. No, it wasn't part of the contract. It's a gift for NASA. I repeat, we are not charging for it, OK? Is it what? Is it dangerous? Oh, for the love of . . . Yeah, it would make your voice squeaky if you breathed it. Please!*

Look, the second purpose of this press conference is to tell you about the other ten tons we're bringing back. OK, I see I've got your attention, finally.

What we have is ten tons of helium-3 extracted from the atmosphere of Uranus during the floater mission. This payload of clean fusion fuel is available for use in commercial fusion reactors starting two weeks from today.

Questions?

Yes, you. Isn't fusion uneconomical? Well, certainly lunar helium-3 has so high a cost that it hasn't brought down the consumer price of electric power as everybody hoped. But we use it because the cost of fossil fuels is going up, their supply is going down, and the environmental effects of burning oil and gas are so severe. But helium-3 is a very clean fuel. What's that? Isn't this a lot of helium-3? Yes, it's a lot. It would power all of North America for over a month. That's why we went to the trouble and expense of bringing it back. Remember, we're over $220 million out of pocket at this point.

Next question. What's it worth? Well, we'll have to see how the bidding goes. At present electric power prices, it would be around $200 billion, but we'd like to see those prices drop. That's right, billion, not million. No, of course we're not ripping off NASA! We won that contract fair and square, and we're delivering more than we promised in the contract! We were the lowest bidders, we took a big risk, we won big, and we pay taxes on our profits . . . enough taxes to buy NASA!

You still don't get it, do you?

—Science Plus

The energy resources of the inner solar system, including both solar power and fusion of lunar helium-3 with terrestrial deuterium, are sufficient to meet Earth's foreseeable energy needs for the indefinite future. But power for activities beyond the asteroid belt and for propulsion of spacecraft is harder to find. Combustion is grossly inadequate for long-term power needs, and fission carries with it the twin curses of weapons-grade fissionables and radioactive waste disposal. The Sun is of course available as a reliable source of power, but the intensity of sunlight is about ten times lower in the belt than in the inner solar system. Solar-power schemes are less attractive in the belt than near the Sun, and solar power quickly becomes impractical beyond the belt, where the intensity of sunlight becomes very feeble. At Saturn, the intensity of sunlight is 100 times weaker than at Earth, making it much harder to achieve economical exploitation of solar power there.

Having learned the benefits of fusion of deuterium with helium-3 during our visit to the Moon, we would certainly like to find larger and more convenient reserves of helium-3 elsewhere in the solar system. An obvious question is whether the surfaces of the asteroids might be rich in helium-3 implanted by the solar wind. On the Moon we saw that the absence of an atmosphere and the feebleness of the lunar magnetic field allowed the solar wind to strike the surface unimpeded, except for "shadowing" of the near side of the Moon by Earth's magnetosphere field near the time of full moon. Implantation of hydrogen and helium and other, rarer elements into mineral grains occurs readily on the Moon, especially in the abundant lunar mineral ilmenite. Asteroids resemble the Moon in that they are too small to retain atmospheres and too cold to maintain the internal dynamo activity that generates magnetic fields. The solar wind strikes their surfaces unimpeded.

But the circumstances of the asteroids differ from those we have studied on the Moon. First, the intensity of the solar wind drops off roughly with the square of the distance from the Sun. Therefore, an asteroid near 3.2 AU from the Sun experiences a solar wind flux that is only a tenth that felt by the Moon. Second, accumulation of solar wind gases on the Moon is aided by "gardening" of the lunar surface by small impacts, which constantly expose fresh material to the solar wind and constantly bury gas-saturated surface grains out of harm's way. On asteroids, even small impact events can remove regolith material by ejecting it at speeds greater than the asteroid's tiny escape velocity. Thus, "mature" surface grains with high contents of implanted gases are preferentially lost, not preserved. Finally, the total exposed surface area of the asteroids is less than the surface area of the Moon. For all these reasons, extraction of helium-3 from the surfaces of asteroids is not likely to be competitive with that from the Moon.

As a planetary scientist who realizes that helium-3 is a highly desirable resource, I am inspired to ask two simple questions: First, where is the helium-3 in the solar system actually to be found? Second (and on a more practical note), what is the best place in the solar system from which to extract and retrieve helium-3 for use on Earth?

The Sun contains more helium-3 than all the other bodies in the solar system combined, but there is no thrill in these words. Getting helium-3 out of the Sun is simply a technical impossibility. It is true that the Sun emits helium-3 in the solar wind, and that in principle we could intercept it. The problem is that the flux of helium-3 atoms is so tiny that it is not practical to collect it in any known way: the grains on the lunar surface that have approached saturation of hydrogen and helium have been exposed for thousands to millions of years—far longer than we can afford to wait.

After the Sun, by far the largest amounts of helium-3 in the solar system are in the giant planets Jupiter, Saturn, Uranus, and Neptune. Jupiter, the largest and closest of the giant planets, orbits more than five times as far from the Sun as Earth. The noontime Sun at Jupiter is twenty-seven times fainter than on Earth. Saturn gets less than 1 percent of the sunlight of Earth, Uranus is even darker and colder, and Neptune has only a thousandth of the intensity of sunlight that we grew up with on warm, sunny Earth.

The giant planets, with such feeble solar heating, are also very cold. Jupiter's face is obscured by a complex series of cloud layers, the topmost of which is a cloud made of tiny crystals of solid ammonia at the bracing temperature of 126 Kelvin (−232 Fahrenheit). The part of the atmosphere we can see, near the cloud tops, is mostly hydrogen and helium gas with a fraction of a percent of methane and ammonia. Some neon and argon is probably present as well. Above the clouds the ammonia is mostly frozen out. Tiny traces of other gases that are made from methane and ammonia by ultraviolet sunlight have also been detected in this region, including acetylene and hydrogen cyanide. Not a very friendly—or healthy—place.

Farther from the Sun, it is even colder. Out at Uranus and Neptune the cloud-top temperatures are 59 Kelvin (–353 Fahrenheit). That close to absolute zero, hardly any gas can avoid condensing or freezing. Methane condenses to form a cloud layer above the ammonia clouds, leaving a gas that is almost entirely hydrogen and helium. Even argon may condense on Uranus and Neptune to form a thin haze above the methane cloud layer. Our experience on the sunny, warm Earth give us no hint of what such temperatures are like: nitrogen and oxygen, which make up 99 percent of Earth's atmosphere, would freeze solid and fall as snow.

So feeble is the Sun as seen by the giant planets that the tiny trickle of heat out of the interiors of these planets is actually larger than the flow of heat from the Sun. The giant planets are, in fact, warm inside. The deeper we look into their atmospheres, the warmer it is. On rare occasions we can see, through breaks in the clouds, into Jupiter's atmosphere down to the level that is about as warm as Earth's surface. Measurements at radio wavelengths show that the temperatures continue to rise with depth. Indeed, the only way there can be a flow of heat out of their interiors is if the atmospheres of the giant planets are much warmer inside than they are at the cloud tops.

A natural question is to ask whether the *surfaces* of these planets might be comfortably warm. But that betrays a terribly parochial notion of what giant planets are like. When we drop a probe into the atmosphere of, say, Jupiter (as the *Galileo* spacecraft mission did in December 1995), our probe measures increasing temperatures and pressures all the way down to the point where the probe is baked by the heat or crushed by the atmospheric pressure. The atmosphere at a depth of a few hundred kilometers below the clouds must have a pressure of thousands of atmospheres and a temperature over 3,000 degrees Celsius, hot enough to boil iron. And still the atmosphere continues. The lower atmosphere is so hot that it is simply impossible to have a solid surface. A witches' brew of chemicals—essentially hydrogen and helium plus all the other ingredients of solar material that are not condensed—compressed to a density close to that of water by the immense pressure, seethes and glows at depths where the Sun has never shone. And even deeper, the pressures are higher. Eventually the pressure gets so high that molecules of hydrogen (each normally made of two hydrogen atoms) get squeezed into permanent contact with each other: the empty space between molecules, which normally is the defining characteristic of a gas, disappears as the molecules are crushed together. In the crush the electrons of these molecules smash together like punks at a dance. The electrons become confused, often losing track of the partner they brought to the dance. As these disoriented electrons begin to rove freely through the crowd of hydrogen nuclei (protons), the fluid becomes an electrical conductor. This incredibly dense, hot soup of protons and electrons is called fluid metallic hydrogen.

At the centers of the giant planets the pressure is many millions of atmospheres and the temperature is far hotter than the surface of the Sun. Slow leakage of heat from the interior tends to deflate the planet, letting it shrink in radius by a tiny amount each year. This shrinkage increases the central temperature and pressure of the planet: heat loss from the top of the atmosphere makes the core hotter! So far as we can tell, Jupiter does not even *have* a surface. The entire planet is fluid, like a star. But there the resemblance ends: many authors refer to Jupiter as a "star that failed," suggesting that there are only small differences between Jupiter and a star like the Sun. This is so gross an exaggeration that even poetic license (for which few scientists have bothered to apply) does not excuse it. Jupiter is about *fifty times* too small to maintain the nuclear fires of fusion in its core. The smallest, feeblest stars, called M5 red dwarfs, have about 5 percent of the mass of our Sun. Jupiter's mass is about 0.1 percent of the Sun's.

Chemically, Jupiter and Saturn are much more starlike. Jupiter has about the same overall element abundances as the Sun, except that the elements heavier than helium are slightly enriched (by about a factor of five) compared to the Sun. These "heavy" elements (carbon, nitrogen, oxygen, and so on) still make up only a few percent of the total mass of the planet. In Saturn, the enrichment of heavy elements is about a factor of ten. In Uranus and Neptune, the heavy elements make up more than half of the mass of the planet. Thus Jupiter, overall, is about 95 percent hydrogen and helium, Saturn is about 90 percent hydrogen and helium, and Uranus and Neptune have about 50 percent hydrogen and helium. But in the cold upper atmospheres of all four planets the heavy elements are almost perfectly condensed and precipitated out. The sole known survivor of these enriched heavy elements near the cloud tops of Uranus and Neptune is methane (and probably argon as well, but we cannot detect argon from Earth). The cloud-top atmospheres of these planets are all rather simple in composition, dominated by the uncondensable hydrogen and helium. The concentration of helium-3 in the atmospheres of all these planets is about the same as it is in the Sun's photosphere. Typically, the giant planets contain about forty-five parts per million of helium-3 in their atmospheres. Chemically, all the giant planets are about equally attractive as sources of helium-3.

"Mining" the giant planets for their helium-3 content is challenging not because of the low temperatures (they help) or their vast distances (we have had several missions that have gone farther) but because of the difficulty of getting out of their gravity fields. The massive gravity wells of Jupiter and Saturn make them very unattractive sources of materials. Their escape velocities are about 60 and 36 kilometers per second, respectively, much higher than Earth's 11.2 kilometers per second; but Uranus and Neptune have low enough escape velocities, about 20 kilometers per second, to be amenable to export of helium-3. The trip times to Uranus are at least a few years each way, and Neptune is about ten years each way. In terms of the demands placed on spacecraft propulsion systems, Uranus is easiest and Neptune is second. Saturn is a poor third, and Jupiter is dead last. We simply do not know of any way to get a single-stage rocket out of Jupiter's deep gravity well and on its way back to Earth.

Of all the Jovian planets, then, the target of choice for helium-3 retrieval is Uranus. It is possible to sketch out a way in which we might plan on retrieving helium-3 from Uranus. We would begin by launching a Uranus entry probe from a space station in orbit around Earth. The probe would perhaps be "slingshot" through Jupiter's or Saturn's gravity well to cut down the trip time to Uranus. After about six years of flight, the probe enters the upper atmosphere of Uranus at high speed. A heat shield, very similar to that carried by the *Galileo* entry probe, protects the probe mechanism during entry and deceleration to below the speed of sound. Once the probe drops to subsonic speed, it deploys a small parachute to drag out a larger chute, which in turn slows the probe down to low speeds. Then a balloon, looking like a bundle of long cylinders, is unreeled from the probe and inflated with atmospheric gases. The heat shield and parachutes, no longer needed, are dropped off, and a megawatt nuclear heat source starts up and begins heating air and flushing the balloon cluster with heated air. The probe slows in its descent as it passes through ever denser layers of atmosphere until, at a pressure of a few atmospheres, it finally reaches neutral buoyancy and stops falling.

Pumps and refrigeration equipment, powered by the heat source, turn on and start liquefying the atmosphere. Any liquid methane is dumped overboard (after using it to help cool the incoming gas stream). Pure hydrogen gas rejected by the helium extraction process is used to flush out the balloons yet again to further increase their buoyancy. Liquid hydrogen is used to precool the incoming gas stream before being dumped overboard, and only helium is retained. The helium is liquefied, and its light isotope (helium-3) is separated from the much more abundant helium-4 by taking advantage of the very different behavior of these liquids at temperatures close to absolute zero. The helium-4 is dumped overboard after using it to precool the incoming helium stream. The remaining helium-3, worth (in 2030 A.D.) about $16,000,000 a kilogram ($7,400,000 per pound), is stored for return to Earth. Waste heat from the refrigeration process (like the hot air that rises out from behind your refrigerator) is used to keep the hot-hydrogen balloon buoyant.

As the helium-3 tank fills up, the mass of the probe increases a little, but not enough to cause a problem with its buoyancy. Finally, when the payload tank is full (our mission returns ten tons of helium-3), we need to make arrangements for our return to Earth. We then start to accumulate liquid hydrogen in an empty fuel tank. Since most of the mass of Uranus air is hydrogen, most of the output from the refrigeration system is liquid hydrogen. Filling the hydrogen tank is a relatively quick process. The waste heat from hydrogen condensation is dumped into the balloons to raise their temperature and increase their buoyancy to compensate for the increased mass of liquid hydrogen in the tank.

When the liquid hydrogen tank is full, we turn on our nuclear rocket engine until the particle-bed or ribbon-fuel core of the reactor is white-hot, then spray liquid hydrogen through it. We would require an engine with close to a million pounds of thrust, and a power of about 20,000 megawatts. This sounds enormous, but the ten-ton Phoebus 2A test engine, with a quarter million pounds of thrust and a power output of 5,000 megawatts, was designed in 1965, built in less than two years, and tested in 1968! I assume that six decades of technological advances will permit an improvement in the power output from about 0.5 kilowatts per kilogram (in Phoebus 2A) to about 2 kilowatts per kilogram in 2030. The hydrogen flash-boils and warms to several thousand degrees, then exits at a speed of nearly 12 kilometers per second through a rocket nozzle. As soon as the engine thrust is high enough to lift the vehicle, it lets go of the balloon and processing equipment and rockets into the sky riding its exhaust plume of fiercely hot hydrogen gas. The engine firing will send the payload coasting all the way back to intercept Earth on the opposite side of the Sun, arriving about six years after leaving Uranus.

What do we do with our ten tons of helium-3 when we get back to Earth? The market value of that amount of helium-3 is set by the amount of energy that it can produce when used in a helium-3/deuterium fusion reactor. That cash value is $160,000,000,000. (Platinum sells for about $460 per troy ounce, which is $15,000 per kilogram. That means that helium-3, at $16,000,000 per kilogram, is *worth 1,000 times its weight in gold or platinum.*) Here is surely the most valuable raw material in the solar system, well worth the cost of transportation back to Earth!

What about the energy cost of retrieving the helium-3? Do we actually come out ahead by carrying out this scheme, trading uranium energy for helium-3 energy? The answer is reassuring: the energy payback for this process is above 20,000:1.

How much helium-3 would we need to power the industrialized nations of the world? The present power consumption of Earth is 8,500 gigawatts, equivalent to fusing 450 tons of helium-3 per year. If we allow for modest further population growth and for raising the standard of living of all the people of Earth to that found in the leading industrial nations at present, we can anticipate using up to 80,000 gigawatts (80 terawatts) of power, for a consumption of 4,000 tons of helium-3 per year—billions of tons of coal or barrels of crude oil. This level of consumption would require one 10-ton load of helium-3 per day, or three 100-ton loads per month. The market value of that much power at present prices is $64 trillion per year.

How much helium-3 is there? If we consider only the outer shell of atmospheric gases on Uranus, down to the level where the pressure is twelve atmospheres, the total amount of helium-3 now present is 16×10^{18} grams, or sixteen trillion tons. This is sixteen million times as much as the estimated amount of helium-3 on the Moon. At the consumption rate of 4,000 tons per year, this would represent a *four-billion-year supply of energy for Earth.* It is enough to keep us going until the Sun runs short of its hydrogen fuel, inflates into its red-giant phase, and swallows up the terrestrial planets.

But the *total* amount of helium-3 in Uranus and Neptune is vastly larger than this paltry sum. The total in those two planets is 5×10^{24} grams. And we know that Jupiter and Saturn have vastly more helium-3, enough to support a population several million times as large as that of Earth for the entire future of the solar system. Getting the helium-3 out of Jupiter and Saturn is more difficult, since neither chemical nor nuclear thermal propulsion is adequate to escape from their powerful gravitational fields. But, if we really needed it, we could think of some way to get it out.

So far we have concentrated on the use of helium-3 as a power supply for planet-bound civilization. But mastery of helium-3/deuterium fusion may permit other entirely new applications. The most significant of these may be the use of helium-3 in spacecraft propulsion, in a fusion rocket. Of the available choices for fusion energy, helium-3 fusion with deuterium is the one that produces the smallest number of neutrons and generates the least amount of radioactive waste. J. F. Santorius, of the Fusion Research Institute of the University of Wisconsin, has presented a very strong case for the development of fusion propulsion systems for space vehicles. He has examined a number of competing fusion schemes, and found helium-3/deuterium fusion to be the most attractive. The most popular fusion reaction for use on Earth, the fusion of pairs of deuterium nuclei, reacts in two different ways. One of these produces abundant neutrons that not only waste 75 percent of the energy liberated by the reaction but also induce intense radioactivity in the reactor walls. The other produces one radioactive tritium nucleus for each fusion event. The competing deuterium/tritium fusion scheme requires intensely radioactive tritium as a starting material. Since tritium has no natural source that is even remotely capable of providing the required amounts for commercial energy production, huge and expensive tritium factories must be built to produce thousands of tons of this extremely dangerous gas. Once the tritium is made and used in a fusion reactor, neutrons are produced in huge quantities, carrying off and wasting some 80 percent of the total fusion energy. Other competing reactions, ones involving boron and lithium, are "clean," but they suffer from much lower reaction rates and much higher ignition temperatures.

The principle of helium-3/deuterium propulsion can be stated in deceptively simple terms: a fusion reactor collides hot helium-3 and deuterium nuclei at high energies. The products are fast-moving protons and helium-4 nuclei. An intense beam of very energetic protons and helium-4 nuclei, carrying all the energy liberated by the fusion reaction, jets from the rear of the rocket to provide thrust. Two interesting options are available: First, any other material may be added to the exhaust stream to "beef it up" and improve the thrust level. Any industrial byproducts or waste can be used for this purpose. Second, the intense beam of positive ions can be used with very high efficiency to generate electrical currents for use on the spacecraft. This *fusion rocket* has such a high exhaust velocity that it can outperform the best possible chemical propulsion systems, like hydrogen/oxygen rockets, by enormous factors.

The fusion rocket could, for example, lift heavy payloads out of the atmospheres of Jupiter and Saturn—such as big tanks of helium-3. Thus, helium-3/deuterium fusion is not simply an end in itself but also an extraordinarily useful tool.

What else could we do in the solar system with helium-3/deuterium propulsion? Santorius points out a few of the immediate benefits. First, trip times for fast flights to Mars can be dropped to as little as two months, or fusion-powered outbound trips to Mars can be flown in eight months with a fuel mass less than 20 percent of the payload mass. Comparable trips to the heart of the asteroid belt would take about eighteen months. Cargo deliveries from Uranus and Neptune could be speeded up from six years to about two. Even manned missions to Jupiter (as in Arthur C. Clarke's *2001: A Space Odyssey*) could be done with trip times of eighteen months to three years. Manned flight to any planet, asteroid, satellite, or comet in the solar system is within our reach if we master helium-3/deuterium fusion. *The solar system is ours to take.*

14

ENVIRONMENTAL SOLUTIONS
FOR EARTH

Jimbo, brewski in hand, watched the wild press conference on TV. "Hey, Janie, listen to this. Some weird scientist talking about billions and billions. Money from your anus or some crazy thing. It's a wonder what kind of crap they let them show on TV nowadays."

The power cord of Jimbo's high-resolution TV screen ran through the wall to a breaker panel in the carport. The power drop drank from a power main that ran out into the Arizona desert south of Gila Bend, where a thousand-hectare antenna received a microwave power beam from an SPS satellite in geosynchronous orbit. Several square kilometers of solar cells on SPS 23 collected sunlight and converted it to microwave power for transmission down to Earth. SPS 23, like its fifty-odd companions, was built mostly of iron and silicon from the near-Earth asteroid 2008 DS. A receiving antenna on the backside of SPS 23 aimed perpetually at the Moon, where three vast solar-cell farms girdling the lunar equator and a large fusion generator in Mare Serenitatis, burning lunar helium-3 with terrestrial deuterium, poured power down to Earth in an invisible torrent. Jimbo had heard most of this, but he found it complicated and therefore confusing, and therefore best avoided.

Janie peeked into the den, resting her hand on the door jamb. Her platinum rings and earrings sparkled in the light. The platinum in her jewelry had come from the asteroid Nereus and had cost her $3.75. "Oh, Jimbo, you know I hate science. Science is boring!" She walked over to the window where the afternoon Sun streamed in and looked out *toward the mountains through clean air under a blue sky, where a few decades ago, back when people had to work twenty or forty hours a week, dust from the big open-pit copper mine and haze from the coal-fired power plant had made the sky a stinking gray bowl. "Let's go for a ride and get away from all that stuff."*

Jimbo hopped to another channel, ignoring her. Up came an American League West game between the Tokyo Giants and the Oakland A's. Jimbo, who followed the exploits of the Giants' power hitter Rosy Linares, settled into the game. Seventh inning, one out for the A's, Danny Phillips on second, two and one on Booz Jenkins. The broadcast came into Jimbo's TV from an antenna the size of a dinner platter in his attic, permanently aimed at a direct-broadcast comsat in geosynchronous orbit. The comsat, like the SPS 2000 kilometers away on the same orbit, was built mostly of asteroidal iron and steel, this from 1998 WX, a metallic near-Earth asteroid whose orbit threatened a devastating collision with Earth in 600 years. At present mining rates, the threatening asteroid would be used up long before it was due to hit. Aboard the comsat, a space-age computer monitored Jimbo's and Janie's entertainment preferences and tailored their programming to their every whim. The computer deemed it important to expose Jimbo and Janie to at least ten minutes of news every day, and it carefully selected what advertisements to show and when to show them. It wasn't like the old days, when people had had to make difficult decisions for themselves.

Janie retreated to the living room, still interested in going out for a drive, and called up the wall TV. The game was still on, but nearly over. She called out "Window 2, weather," and a live weather satellite view of her region popped up. "Animate today," she commanded, and her screen ran a video of the weather patterns in her area for the previous twelve hours, pulled off a Russian metsat in polar, Sun-synchronous orbit, supplemented by a supercomputer projection of the weather of the coming twelve hours for her region, downloaded via satellite relay from a mainframe computer in Beltsville, Maryland. The display image was marked by several simple icons in bright, primary colors. A synthetic custom voice-over from their service computer in GEO gave the details for their precise location. Janie was vaguely aware that things used to be different, back when people had to read and write. That whole concept seemed terribly difficult and unbearably tedious.

"End window 2; sound off window 1; notify break window 1." The home computer, based on gallium arsenide VLSI chips made in a microgravity factory in Space Station Gamma's industrial park, would now beep to warn her of the end of the game. She went back to her painting: a scene of Elvis and Jackie, surrounded by stylized signs of the zodiac, in white and red and silver-glitter paint on black velvet.

A few minutes later the household computer discretely beeped. Janie walked into the kitchen, gave the sink a pile of brushes to clean, and intercepted Jimbo at the den door. "How about that ride, big boy?"

Satisfied by the outcome of the game (6–4 Giants), Jimbo took her arm and walked her to the door. "Leaving; lock all; security system set." They exited to the carport and hopped into their car. "Start." Jimbo then recited a short voice-recognition litany.

"Where to, Jimbo?" asked Mario, their driver.

"To the overlook on Wildcat Mountain, Mario. Park in the picnic area. Oh, call Annie's Deli and order picnic basket number four. We can pick it up on the way."

Mario's algorithms unscrambled the logic, calculated their ETA at Annie's, called in the order, and got under way. Using the Improved Civil Protocol, Mario tuned into the EGPS satellite navigation system and tracked his position and progress with a precision of a few centimeters, while watching and allowing for the sparse traffic and rare road hazards with his on-board sensors. It was all perfectly safe, of course, not like the old days, when there were accidents and people could get hurt.

The ceramic propane fuel-cell engine poured out electric power silently, the tires humming on the pavement, venting an odorless, nontoxic, smoke-free exhaust cleaned by its rhodium catalytic converter. The rhodium, like the iridium electrical contacts in the fuel cell, had come from the same asteroid as Janie's jewelry.

On the way out of town they passed a billboard that said We Care for You in large block letters. Mario, noticing that they were looking at the billboard, read it out loud to them in his warm artificial voice.

"Yes," thought Jimbo and Janie quite spontaneously and independently, "this is the way things were meant to be."

—Stay-at-homes

Earth in the coming third millennium faces a variety of urgent environmental problems. The three central problems are energy, mineral resources, and food. We shall look at the tradeoffs presently facing decision makers in each of these areas, but we shall not attempt to predict what decisions will be made. Predicting future events is an imperfect art at best. I am reminded of the great baseball philosopher Yogi Berra's immortal dictum that "The trouble with predicting the future is that it is very hard." In that spirit, we shall begin with a discussion of Earth's energy problem.

Energy is the primary requirement of modern civilization. Per capita energy consumption is well correlated with per capita wealth, a linkage that can be weakened (but not removed) by careful attention to energy efficiency and recycling. Expensive energy makes the production, transportation, and refrigeration of food more expensive. Rising food prices have a disproportionate impact on the poor, who have the fewest options for rebudgeting money from other categories.

Millions of informed people are concerned about future energy supplies for Earth. The dwindling supplies of crude oil and natural gas are frequently discussed in newspaper articles. The immediate alternative, the accelerated use of nuclear power, faces determined opposition from well-organized environmental groups throughout the Western world, fired by the disturbing facts about reactor safety in power plants built by the former Soviet Union. The only other significant source of power today is hydroelectric. Very small amounts of power are also produced by solar, geothermal, wind, and tide energy, all of which face serious obstacles of availability and economics. In less-developed nations, power from burning biomass is also important locally. Earth's increasingly desperate need for new sources of abundant, inexpensive, clean electrical power has stimulated much discussion of alternative energy sources.

At present, the largest single source of power in the world is fossil fuel combustion. The sources of these fossil fuels are hard (anthracite) coal, soft (bituminous) coal, crude oil, and natural gas. Vast but dilute fossil fuel reserves are also found in oil shales and tar sands, few of which can be worked economically. (Indeed, in many cases the energy required to extract the fuels would be greater than the energy content of the fossil fuels extracted!) Huge reserves of as-yet untapped gas hydrates underlie extensive areas of Siberia, Alaska, Canada, and the sea floor. At present, the lion's share of global energy derives from the vast crude oil reserves of the Middle East, although large amounts are also produced in the United States, Venezuela, Russia, and several other nations.

The known reserves of coal undoubtedly are sufficient to last for centuries at present consumption rates. Unfortunately, some developing areas of the world, such as China and Eastern Europe, must base their plans for industrial development on their principal native energy resource, bituminous coal. Soft coal is notorious for its content of pollutants, such as sulfur. Combustion of sulfur-bearing coal in Poland, the former German Democratic Republic, the Czech Republic, and Russia has released vast quantities of sulfur dioxide into the lower atmosphere, where further oxidation and hydration produce sulfuric acid. The result has been the generation of terrible smog, acid fog, and acid rain in quantities sufficient to toxify and kill entire river systems and make lung diseases pandemic.

A further, little-noted hazard of some coal deposits is the peculiar geochemical transport of uranium and thorium by groundwater into coal veins. The strongly reducing conditions maintained by reactive organic matter in coal causes minerals containing low oxidation states of both of these radioactive elements to precipitate in veins in the coal. These yellowish, powdery deposits—called thucolite after the chemical symbols of thorium (Th), uranium (U), and carbon (C)—are released as a fine ash when the coal is burned. Fly ash from combustion of thucolite-bearing coal is intensely radioactive; indeed, electrical production by combustion of such coal releases more radioactivity into the environment than a comparable-capacity nuclear power plant. Further, soft coal sometimes contains dangerous concentrations of toxic materials such as arsenic, antimony, and cadmium. It has become a common problem in China for a family to cook a meal over bootlegged or scavenged soft coal and suffer serious illness or death from arsenic poisoning.

Another large reservoir of fossil fuels, solid gas hydrates, has recently come to public recognition. A variety of evidence suggests that large areas of permafrost at high latitudes may have been converted to solid gas hydrates by natural gas percolating upward from gas domes in the crust. Expanses of these gas hydrates, still poorly mapped, have been reported in Siberia, the Canadian Arctic, and Alaska. More recently, evidence has appeared that vast quantities of solid gas hydrates are formed in ocean-floor sediments by the upward diffusion of organic gases from the sediments below. Estimates of the size of this source of natural gas are still very indefinite, but some experts believe that the total energy content of these gas hydrates may surpass the total supply of extractable crude oil several times over.

But all fossil fuels, especially the cleanest anthracite, burn to produce abundant carbon dioxide. This gas is a powerful contributor to the greenhouse effect, helping to make the atmosphere opaque at wavelengths between the major infrared absorption bands of water vapor. This enhancement of the greenhouse effect leads to increasing Earth-surface temperatures and global climate change. Global warming directly causes melting of polar ice, which raises sea levels worldwide by tens of meters, inundating seaside cities and invading rich cropland. The evidence linking carbon dioxide emission to global warming in this century is far from convincing and certainly contains features that are not explained by any theory. But the general principle seems secure: we must beware of any further large increases in the carbon dioxide level.

An obvious and technically achievable alternative to fossil fuel combustion is nuclear fission. Nuclear power is derived from uranium, which is mined in Canada, the United States, the Congo, Russia, Australia, and several other places widely but unevenly distributed throughout the world. The isotope U-235 is unstable, decaying by a process called spontaneous fission. A single U-235 nucleus has a 50 percent chance of spontaneously splitting within 1.4 billion years. Very large ensembles of U-235 nuclei obey a very precise decay law, in which half of the nuclei decay in 1.4 billion years, half of the survivors decay in the next 1.4 billion years, and so on. The time interval over which half of the nuclei decay is the *half life* of U-235. Each U-235 nucleus that decays spontaneously emits two large but unequal fragments, plus several neutrons. Some of the fission fragments are themselves radioactive. A large amount of energy is given off during spontaneous fission, both in the form of high-speed fission fragments and direct nuclear radiation, such as gamma rays. In a small piece of U-235, the neutrons given off by decay generally escape without causing further reactions. But in a large, pure mass of U-235 the neutrons given off by the decay of one atom may collide violently with other nuclei, causing them in turn to split. This is the famous nuclear chain reaction. When the neutrons from one fission event produce exactly one further neutron-induced fission event, this is a stable chain reaction: we say that the reaction has "gone critical." Devices that are designed to produce stable chain reactions are called nuclear reactors.

Nuclear reactors may be designed for power production. In such power reactors, the enormous amount of energy given off by nuclear reactions is used to, for example, boil water and turn steam turbines, which in turn drive electrical generators. Alternatively, the reactor may be used as a source of neutrons to convert nonfissionable U-238 into a fissionable material, with a resulting large gain in the total mass of fissionable material. The product is plutonium, an artificial element whose existence in nature was unknown at the time it was first made in the laboratory. The reactor type used to make large amounts of new fissionables is called a breeder reactor. Plutonium is an exceptionally useful element for making small, compact nuclear weapons and is in great demand among terrorists.

Neutrons from any nuclear reactor can be used, either intentionally or inadvertently, to make heavy isotopes of almost any element. Many of the isotopes made by this process of neutron bombardment are radioactive. Radioactivity is induced in the metallic containment vessel that surrounds a reactor core by neutrons that escape from the core. Both radioactive fission products and induced radioactivity in structural materials contribute to the problem of radioactive waste.

A mass of U-235 slightly larger than that required for a stable nuclear reaction (a supercritical mass) will trap a slightly larger fraction of its own emitted neutrons, which will cause the rate of induced fission—and the rate of energy production—to accelerate steadily. The mass of uranium soon becomes so hot that it melts and disperses, a phenomenon called meltdown.

If a highly supercritical fissionable core of material is assembled very quickly and held together for a matter of microseconds (by firing the pieces of the mass together in a gun, or squeezing a small, slightly subcritical sphere very hard by detonating a shell of chemical explosives surrounding it), we get a rapid chain reaction that liberates vast amounts of energy almost instantaneously. This is commonly called an atomic bomb, although that name is a little silly, since all bombs, whether chemical or nuclear, are made of atoms. A much more cogent name is "nuclear explosive" or "fission bomb." The fissioning of a kilogram of plutonium, roughly the size of a golf ball, gives off about as much energy as one thousand tons of chemical high explosives. We would call this a one-kiloton nuclear explosion. The device that produces a kiloton explosion can be packaged inside an artillery shell— or a briefcase.

Of course, some negative environmental impact results from the mining of uranium, including not only the usual unsightly scars of mining but also stream pollution by radioactive dust. In general, however, these problems pale to insignificance compared to the problem of radioactive waste disposal.

There are also valid environmental concerns about reactor safety, brought home to us all by the disturbing saga of the Chernobyl reactor in the Ukraine. Astonishingly, several old Chernobyl-type reactors are still in service, despite their well-known design defects. But nuclear engineers have known for many years how to design a safe reactor. Discussions of future reactor safety should revolve about two critical issues: nuclear waste disposal and nuclear weapons proliferation. These two central problems associated with nuclear power can be ignored only at our peril.

The desire to minimize the environmental impact of emerging nations while greatly increasing their standard of living (that is, greatly increasing their per capita energy production), has led to the attempt by many developing nations, including India, Pakistan, Iran, Iraq, and many others, to develop nuclear power reactors. But the technology that makes power reactors possible also makes breeder reactors feasible. Thus, there are serious grounds for linking nuclear power production with the potential for proliferation of nuclear weapons. The established major nuclear powers, the United States, the former Soviet Union, England, France, and China—are generally resolved to minimize the proliferation of nuclear-weapons technology. The most effective single means of doing so is to refuse technical, economic, and material assistance to other nations for the construction of breeder reactors. Despite this general resolve, it is impossible to "uninvent" a technology that is already widely available. The problem of who should "police" noncompliant nations also raises major ethical and political dilemmas.

Despite the intent of the nuclear powers to prevent nuclear weapons proliferation by controlling the export of technology and materials, many nations either have or will soon have the ability to manufacture their own nuclear weapons. These include Israel, Iran, Iraq, Japan, India, Pakistan, South Africa, Australia, Sweden, Italy, Brazil, and Egypt. For normal wartime military purposes, some means of delivering these weapons to their targets are required. Several of these nations have either acquired short-range nuclear-capable missile systems from the former Soviet Union (the Scud missile) or are developing rocketry on their own. But for terrorist uses, no such delivery system is needed. Any nation that has a breeder reactor and supports terrorist activities can threaten any city on Earth. The only effective countermeasure to such activities is international inspection of all the nuclear facilities on Earth. But this is in turn dependent on every nation granting free access to inspectors from international regulatory agencies to all of their most sensitive facilities, both nuclear and nonnuclear: an inspector cannot confirm that forbidden breeder technology is absent from any site unless that site is open for inspection. Governments planning for war or consumed by paranoid (or entirely valid!) suspicions of their neighbors will not willingly grant such access, as the behavior of Iraq and North Korea clearly show. The triple combination of nuclear weapons, missile delivery systems for them, and violent or paranoid political and religious control is a recipe for disaster.

Thus, humans' heavy dependence on present sources of electric power has many undesirable aspects. The rational desire to reduce reliance on fossil fuels and uranium means that either an economically competitive alternative source of power must be found or global energy usage must be dramatically curtailed. The latter course, argued by some extremist environmental groups, would clearly result in a substantial decrease in the ability of Earth to grow and transport food. The carrying capacity of our planet would drop to about one to two billion people, the large majority of whom would have to be employed in low-energy, labor-intensive food production. This vast, dispersed rural workforce would need, and would receive, only the education needed for manual subsistence farming. Thus, the price of choosing this path is the loss of virtually all urban population, the reduction of virtually all humankind to medieval standards of living and backbreaking labor, and the loss of at least three billion human lives. If we recall that the rise of civilization was synonymous with the founding of cities; the rise of literacy, religion, education, the arts, and the sciences; and the development of countless other nonagricultural specialties, then we can succinctly characterize this scenario as the end of civilization on Earth.

When these issues first gained widespread public attention, during the onset of the "energy crisis" in the early 1970s, there was much talk of replacing fossil fuels and uranium with "green technologies" for energy generation. It was supposed that burning wood, trash, and the like would provide for our energy needs. But this qualitatively facile approach is a quantitative disaster: the only way we could do this is if we utterly abandoned over 90 percent of our energy consumption. Curiously, this might not result in any reduction in air pollution: biomass fires are notorious polluters, a realization that led to the banning of leaf and brush burning in most urban areas thirty years ago. Wood, brush, grass, and trash fires release vast amounts of pyrotoxins, soot, partially burned organics, airborne ash, and nitrogen oxides. If we chose the biomass route, we could easily reduce our standard of living (which is tied to per capita energy consumption) by a factor of 10 to 100 while increasing air pollution worldwide.

Still, the environmental imperative is real and valid. Ebbing supplies of fossil fuels, emerging environmental hazards (global warming, acid rain, ozone depletion) associated with combustion of hydrocarbons and coal, and the uncertain prospects for safe disposal of radioactive waste from nuclear power plants combine to compel a search for cheaper and environmentally safer energy sources.

Hydroelectric power has some environmental impact, but no negative global consequences. The excavation, blasting, and mining activities associated with dam construction contribute to local air and water pollution, and both silting behind dams and downstream waterflow changes present serious problems. But the main problem with future expansion of hydroelectric power is that the best sites for power dams over the entire planet have already been exploited. Most remaining potential sites are very remote, have limited or erratic water flow, or have a small water drop height, making them much less attractive economically. The sad truth is that this source of power is already nearly tapped out. In any case, it cannot supply more than a few percent of our present energy needs.

Geothermal energy is of great interest in certain locations where hydrothermal steam is vented into the atmosphere. The average heat flow from Earth's interior is so diffuse that using it economically is impossible. The total geothermal heat flow is 10^{28} ergs per year. This sounds like a lot of energy, and on the whole, it is—about a thousand times as much energy as released by all the earthquakes on the planet, and several times the total amount of energy used by human civilization. But that comes to only 5×10^9 ergs per square meter per year, or a power density of 0.000017 watts per square meter. It takes about a kilowatt per person to maintain a technologically advanced civilization with a high standard of living. Allowing a high efficiency of 20 percent for the conversion of geothermal heat into electric power (no one knows how to do it this well), we would need to collect the entire heat flow over an area of sixty square kilometers to meet the needs of a single person! Obviously, the cost of building and installing a system big enough to support the average family would bankrupt a small nation. There are a handful of places, like Yellowstone National Park or Wairakei, New Zealand, and in the volcanic fields of Iceland, where much higher concentrations of heat flow occur locally. These areas of concentrated heat flow can sometimes provide enough power to run a small town. Unfortunately for the future of hydrothermal power, not only does most of the heat reach the surface in too dilute a form for exploitation, but the places that do have concentrated heat flow are mostly along the midocean ridges under thousands of meters of water, impractically remote from centers of population. Geothermal energy is, basically, far too rare and too low-grade for widespread economic use.

Many readers will also be familiar with demonstrations of production of electric power by tapping the energy of winds and tides. However, there are few places on Earth where the winds are both strong and reliable, and people rarely choose to live there. Kerguelen Island in the southern Indian Ocean has strong, reliable winds—and no permanent residents. But the reliability and efficiency of wind-power generators has steadily improved in recent years, suggesting a brighter future for this very low impact technology. The economics of wind power have already been made competitive with fossil fuel burning, and the environmental impact of using wind power is incomparably lighter than that of even the cleanest oil- or coal-fired power plant.

Tidal power faces similar challenges. Tidal ranges are rarely large enough to provide an attractive source of power. The world's highest tides, in the Bay of Fundy, could be tapped by building an immense dam across the opening of the bay. The cost of this endeavor is so enormous that it seems most unlikely to happen. Also, tidal power generators must be designed to withstand severe wave action and the corrosive effects of seawater. Both of these factors drive costs up.

One very interesting emerging technology is the generation of electric power from sunlight using photovoltaic cells. Such solar cells are already in widespread specialty use: my vacations have been made more flexible and independent since 1986, when I bought a square meter of solar cells that generate a total of seventy-five watts of output power in direct sunlight to recharge the storage batteries in my travel trailer. Commercial cells now available on the market can operate at efficiencies of 15 percent to 20 percent, and some experimental double-layer cells have demonstrated laboratory efficiencies up to 30 percent. At Earth's surface the Sun delivers about a thousand watts per square meter when the air is perfectly clear and the Sun is directly overhead. In a typical moderately sunny site at middle latitudes, such an array can deliver the equivalent of five to six hours of noontime-Sun power each day. The overall performance of the system is then about forty to fifty watts per square meter, averaged over day and night.

A solar cell panel repays the electric power needed to manufacture it in its first few hours of operation. Thereafter, the power is produced absolutely free of any form of pollution.

It is easy enough to envision vast expanses of desert covered by "energy farms" of solar cell arrays. With 20 percent efficiency, exposure of 50,000 square kilometers of cells to the Arizona or Saudi noontime Sun would produce 10,000 gigawatts of electric power, well in excess of present world power demands. A single circular array with a radius of about 130 kilometers (85 miles) would power the entire world—as long as the Sun remained at the zenith. Actually, it would be folly to use only a single collector: a cloudy day at one location would shut down the planet. An array in Arizona, another in Chile's Atacama Desert, another in the Sahara, another in the Rub al Khali, another in the Gobi Desert, and one in the Australian Outback would serve the world better—if there were a global power grid to spread the power around. But the Sun does not shine all the time: even in perfectly cloudless weather, the Sun still sets at night. In fact, even when the Sun is low in the sky, there is a serious loss of power, caused by absorption and scattering of sunlight in Earth's atmosphere. One can scarcely expect to average more than about eight hours per day even in the very best locations on Earth. Energy storage, "saving against a run of cloudy days," presents a giant problem. Such energy storage might be most efficiently accomplished by a terrestrial application of fuel cell/electrolysis cell technology developed originally for use in space. Also, the sunniest areas tend to be dusty. The environmental impact of solar power is generally benign, except for local heating caused by the fact that the solar cells are generally darker than local rocks. But this factor can be completely offset by the application of crushed white rock to the ground around the solar cell array, to adjust the overall reflectivity to be the same as it was before the solar cells were installed.

Solar cells in space can, as we have seen, provide twenty-four hours of exposure to a solar-energy flux of 1,350 watts per square meter. At 20 percent efficiency, solar photovoltaic cells in geosynchronous orbit can deliver 270 watts of electric power per square meter continuously.

An identical array on the surface of the Moon is in daylight 50 percent of the time, but it sees the daytime Sun without any atmospheric absorption or scattering. A lunar power station therefore provides an average of 135 watts per square meter of electric power. Both space-derived solar power sources SPSs and lunar power stations easily outperform Earth-surface solar cell farms. And, in the absence of air, dust storms are not a problem on the Moon or in space!

Another exciting prospect for future power generation is terrestrial fusion power. Fusion is the direct combination of the nuclei of the light elements, most notably hydrogen, to make heavier elements and liberate vast amounts of energy. In order to get these light nuclei to approach each other close enough to react, it is necessary to give the colliding nuclei enough energy to overcome their electrostatic repulsion. This requires reactor temperatures of about 100 million degrees.

There are a number of possible choices for the fuel to be used in a fusion reactor. Typically, the fuels are isotopes of the three lightest elements, hydrogen, helium, and lithium. Natural hydrogen (as in seawater) is overwhelmingly composed of the light isotope hydrogen-1. There is also one atom of stable heavy hydrogen, hydrogen-2 (normally called deuterium) for every thirty thousand atoms of light hydrogen. The third isotope of hydrogen, hydrogen-3 or tritium, is highly radioactive and has a very short half-life. It is so rare in nature that it cannot be "mined" or extracted in useful quantities. Thus, any fusion reactor that requires tritium fuel also requires a "tritium factory" to make the fuel, as discussed in chapter 13. On Earth at present, the largest source of tritium is just such "tritium factories" that make tritium for use in fusion ("hydrogen") bombs. Small amounts of tritium are added to the lithium and deuterium in these bombs to speed up fusion reactions in them and make them more efficient. Since tritium has a half-life of only a few years, it cannot be stored for long periods of time. Indeed, fusion bombs and warheads must be periodically disassembled and recharged with fresh tritium.

The light isotope of helium, helium-3, is also useful as a fusion fuel, but it, like tritium, is extremely rare on Earth. The main source of helium-3 is in fact hydrogen bomb recycling! Even that source is far too small to generate any significant amount of electric power. We have earlier seen that the fusion of helium-3 with terrestrial deuterium from seawater is a potential source of clean power: this particular fusion reaction does not release neutrons and produces no radioactive byproducts. Also, since this reaction does not use tritium, it does not require the manufacture or transportation of this dangerous substance. As argued earlier, the only plausible sources of helium-3 are the Moon and the giant planets: they depend on space technology to make clean fusion fuel available. In general, considering the potential of geosynchronous solar power satellites made from asteroidal materials, lunar surface power stations made from lunar materials, and helium-3 return from the Moon and eventually from the Jovian planets, the use of space technology offers the most attractive solutions to Earth's energy problem now visible.

Mining of metals and other industrial raw materials is second only to energy-mining (of coal, crude oil, natural gas, oil shales, uranium, and so on) in its disturbing impact on the biosphere. As industry modernizes and recycling becomes the norm, advanced industrialized nations find themselves meeting more of their needs of industrial materials from recycling and less from mining. Since the energy needed to recycle a ton of steel or copper is far less than the energy cost of mining and processing a ton of new metal from ores, the amount of energy used in manufacturing and the amount of mining of both metal ores and energy resources are now declining in the industrialized world. Stabilization of population in the most advanced countries means that the total amount of metals and other industrial raw materials in circulation on Earth should stop growing and become constant, which in turn means that all current demand could in principle be met by recycling. Primary metal production would become a very minor business.

One extremely valuable scientific concept—the *steady state*—describes this principle. A steady state is one in which certain quantities, such as population or food production rate, are kept constant by means of a steady flow of energy and materials through the system. In a world with constant population and an unchanging standard of living, the amount of each commodity (iron, glass, aluminum, cobalt, platinum, concrete, and so on) in circulation at any time is constant. This means that mining and new production of industrial and commercial raw materials essentially disappear: all mineral needs can be met via recycling—assuming only that an adequate energy supply is available to do the recycling. The rate of power consumption would be quite modest in such a *postindustrial, steady-state* society. The advanced industrial nations are already approaching such a state: population growth is arrested or dwindling, per capita GNP is high and steady, primary mining and production are fading away, per capita energy consumption is declining, and recycling is a large and growing industry. In such a society there is less work to be done than in our present slowly growing economy, but at least the same per capita wealth. The conversion of our economy is not yet complete: the designing of products to maximize the ease of recycling still receives inadequate attention, but this seems to be an inevitable development. The principal challenge of such a steady-state society is to wean itself of energy mining. This requires the development of alternative sources of energy that are either renewable or inexhaustible.

Materials science provides yet another valuable principle: new scientific knowledge and new technologies based on that knowledge constantly unfold new types of materials and new uses for old ones. A century ago aluminum was a rare, light metal that cost more than silver, plastics were unheard-of, and germanium and uranium had no known uses. With increasing knowledge, each raw material takes on ever more possible uses; at the same time, fewer and fewer materials have truly unique uses that could not be met by some other material. This is called the principle of substitution.

A related economic principle is also at work in conjunction with the principle of substitution. Let us suppose that someone discovers that incandescent lightbulbs become 10 percent more efficient and last 25 percent longer if we add five cents' worth of mysterium sulfate coating to the filament. Immediately the demand for the rare element mysterium grows from a few tons per year to ten thousand tons per year. There is so little mysterium production in the world that prices soar: market manipulation by a few traders creates an artificial shortage that runs prices up even higher. Mysterium, available only as a minor byproduct of didymium mines, is never mined for its own sake, and there is no economic justification for vastly increasing didymium production just to make a little more mysterium. Immediately, geologists and prospectors, encouraged by the booming mysterium market, start looking for potential new mysterium mines. At the same time, materials scientists launch an extensive search for other materials that might have similar effects. They soon find that glucinium phosphate, a material available as a byproduct of fertilizer production and having no other known uses, works almost as well as mysterium sulfate and costs a tenth as much. *Research makes new resources.*

So far we have concentrated on the future of the developed nations. But most of the nonindustrialized nations lag far behind the industrialized world in their standards of living and their levels of energy consumption. Their economic fate is an important factor in the future political stability of the world. Will these peoples continue to live in poverty and disease, or will they be brought up to modern standards of living? If the latter, where will the money, resources, and energy come from to effect that change, to provide them with the mineral and energy resources needed to reach a sustainable steady state? What impact will their increased mining and energy production have on the world?

For those who stay home on Earth, few of their mineral needs can be met by mines in space: transportation costs are so high that only intrinsically valuable materials can be economically imported to Earth. Among the few material commodities that are worth importing is the class of precious and strategic minerals. Jeffrey S. Kargel, a geologist at the U.S. Geological Survey in Flagstaff, Arizona, has analyzed the impact of importation of asteroidal platinum-group metals and other valuable minerals on market prices. He concludes that prices of many such commodities can be brought down substantially while still turning a profit. A few of Earth's large mines, such as the Homestake complex in the United States, the Norilsk mines in Siberia, and the deep gold mines of South Africa, would be unable to stand up to such competition—but generally, the total amount of mining being done on Earth would be very little affected by importation of asteroidal precious and strategic metals.

If we really want to reduce the environmental impact of mining activities on Earth, the best place to start is in the energy market. Energy dominates global pollution, through mining, refining, transportation, combustion, and waste disposal. Primary metals production and phosphate fertilizer mining also have major environmental impacts. Reducing mining disruption of Earth's crust is an achievable goal, dependent on our striving for 100 percent recycling. Efficient recycling is the only acceptable solution for Earth and the only key to sustainable life elsewhere in space. Recycling is an intelligence test for humankind that we may choose to win—or lose. Similarly, moving to low-impact energy sources and eradicating the major forms of pollution associated with fossil fuel burning and fission reactors are integral parts of stabilizing Earth's population and economy.

.∴.✳.

The third great agent of environmental disruption, agriculture, is a pervading cause of surface disturbance, erosion, and water pollution by fertilizers, stockyard effluents, fungicides, pesticides, and insecticides. Highly efficient hydroponic agriculture suffers from the high initial cost for construction of facilities. Organic farming techniques promise good sustainable yields, but they are still unfamiliar to almost all commercial farmers. Other issues enter into our choice of future agricultural options: the vast preponderance of farmland and grazing land is used, directly or indirectly, for the extremely inefficient production of meat. Each pound of meat protein delivered to the consumer requires the expenditure of several pounds of plant protein, mostly from field corn, hay, and oats: raising meat *decreases* the amount of human food that our planet can provide. The principal motivation behind massive meat production is economic: there is a huge market for high-priced sources of fat, cholesterol, and artificial hormones. Large-scale meat consumption is as American as smoking or alcohol and drug use: to quote Jimbo on this subject, "Ain't nobody gonna tell me what to do." In Standard English, this means, "If someone presents evidence that what I am doing is stupid and self-destructive, I will punish them by ignoring their advice—even if it kills me." Yet minimizing red meat consumption permits feeding more people from a smaller acreage of farmland and reduces several serious hazards to the public health. Earth could probably support twenty to thirty billion people if farmland were used only to raise food for direct human consumption. If Earth's population continues to grow unchecked, the price of meats will eventually soar, driving them into the luxury market. In densely populated, highly industrialized countries, such as Japan, this trend is already clearly visible.

These lessons will not be lost on the designers of space colonies, for whom growing area is severely limited and efficiency is of the essence. Meat production in space colonies will almost certainly focus on scrap-eating fowl (chicken, turkeys, ducks, and geese), fish, and cellulose-digesting goats. Sheep and some varieties of pigs may also be efficient enough to be worth bringing along. Pigs in space make much more sense than cows in space. Development of this approach in experimental closed ecosystems promises big terrestrial payoffs from this form of space biotechnology.

∴ ∴ ✳

Many people are concerned that some of Earth's problems may be too difficult to solve. Others wonder whether we ought not be trying harder to solve them. These valid concerns have become polarized by political extremists. Much political baggage must be discarded before we can actually get on with the essential business: many simplistic fallacies of the left and right get in our way.

From the left: "Woe is me, we're all doomed. The sooner we decide to die, the better for our planet. Individuals must sacrifice their standard of living (and posterity) for the good of the planet. Let us all take a vow of poverty and chastity right now and stop wasting money on illusory searches for solutions that certainly don't exist. And let us shut down all those fascist industries first. That will help get the population down to manageable levels and let the cute little birdies and snail darters and pink squirrels recover." Many on the far left love trees more than people.

From the right: "There's no problem: industrial productivity is all that matters because it is the basis for individual prosperity. There are no 'rights of the planet' or 'rights of species'—only individual rights. The powers arrogated by government acting in the name of society, such as environmental legislation, must be sacrificed for the good of the individual. Let's stop wasting money subsidizing pinko scientists who want to take our money to study problems that we don't want to know about and don't understand." Many on the far right love weapons more than people.

Once again the human race is taking a continuous intelligence test, as individuals and as a species. We must first define problems (the environmentalists have this part right). Second, we must turn our knowledge and resources toward finding solutions for these problems. And we must look everywhere for those solutions, even to new branches of human knowledge such as space science and space technology. If it turns out that no solution other than mass starvation or death by global pollution is possible, then we are truly stuck. But if we *assume* that our problems are insoluble, we will not try to find solutions, and we can be certain that no solution will be found. *But we are not helpless.* We have a tremendous intellectual resource, people trained in the art of problem solving, to bring to bear on this problem. We need to make it clear that we want and expect solutions from our scientists and engineers, and that we are willing to pay for the solutions now so that we will be alive later.

Extremist rhetoric beclouds what every thinking person already knows to be true: there *are* problems, the problems *must* be dealt with, and solutions *can* be found. Sometimes the only solution we can find will be the renunciation of some kind of human activity, such as the mass production of chlorofluorocarbons. That means the environmentalists were right about something. And sometimes the research will show that federal regulations requiring quarterly analyses of the water in the Santa Cruz River in Tucson (there isn't any water in it!) are crazy. That means the right-wingers were right about something. But the mindless slogans of these groups are still mindless. All we have really proved is that nobody can be counted on to be wrong 100 percent of the time. If some people loudly proclaim that a certain problem is insoluble, we can interpret that to mean they do not know how to solve it. That means it is time to listen to somebody else. So, if we cannot trust political extremists to do our thinking for us, how do we know what is true? By careful, objective, nonpoliticized analysis of the facts.

Many social ills arise directly out of poverty and its associated feelings of disenfranchisement, envy, and alienation. In the postindustrial era, poverty is becoming a hideous anachronism. So is shameless wealth, which feeds on and amplifies the social ills of greed, pride, and oppression. The state of universal moderate affluence, with all the basic material needs of life met, and with adequate time and resources available for education, is sufficient for any psychologically healthy individual. There are not too many affluent people on Earth; there are far too few. Seeking social and economic justice for all is not merely an exercise of compassion: it is good business. *The excuse that there are not enough resources to go around is an obsolete holdover from a philosophy that failed. There is no shortage of material resources.* Seeking affluence for all is possible. It is not merely good business; it is compassionate action. Moral and ethical issues are not optional in making plans for the future of humanity. They are central. The two greatest and most distinctive abilities of humanity, setting us apart from other terrestrial life forms, are intellectual growth and spiritual growth.

One final concept is of great value in understanding wealth: the mathematical principle of the *zero-sum game.* I have found that most people have only a vague perception of mathematicians: they are viewed as remote, abstract, fuzzy-haired dreamers who walk through the snow in bedroom slippers and often wander out into traffic. Civilians are invariably surprised to learn that a genuine branch of mathematics called game theory exists and is populated by a cadre of remote, abstract, fuzzy-haired practitioners called game theorists. These gentle folk study strategies for card games, ways to beat casinos, and similar esoteric disciplines. They enjoy logic games and are prone to utter statements such as "All generalities are false, including this one."

Game theory includes a large set of games in which a certain, fixed number of points (dollars, chips, markers) simply get redistributed during the game. The game starts with so many and ends with the same amount—it just redistributes the wealth. This is a zero-sum game. Casino games are mostly not zero-sum games: they are *decreasing-sum* games, in which a percentage of the wagers is raked off by the casino and disappears from the pot of available winnings. Examples include slot machines, whose long-term players are virtually guaranteed to lose. Casinos love these games because they are the only guaranteed winners. In the wider realm of society, decreasing-sum games are a favorite of centralized, entrenched power structures—at least among those who profit from the house rakeoff.

In a world having no research, no new materials, no new technologies, no new products, and no newly created wealth, the economy is a zero-sum game. In such a world, business arguably is a decreasing-sum game, because the government always helps itself to a percentage of each person's income. Market share and profit margin are everything; no one spends a cent on research or product development. It is a world designed for MBAs *only*.

Now comes the point: *The real world is neither a decreasing-sum game nor a zero-sum game—it is an increasing-sum game.* The total amount of wealth on Earth can and does increase with time as a result of scientific discovery, engineering innovation, and the marketing of new and improved products. It is possible for a group of players to play such an increasing-sum game for a long time and break even, which can be done in a zero-sum game. It is possible for a group of players to play an increasing-sum game for a long time and have many of them come out ahead, which can also happen in a zero-sum game. But it is also possible for *every player to come out ahead.* Such an outcome, possible only in increasing-sum games, is a deadly threat to any individual or organization that desires to exercise control over others. This is why tyrannies of the left and right not only prosper on widespread repression, poverty, and ignorance but also habitually purge anyone with new ideas. And this is why those who flee Earth will be the exceptionally creative.

But resources from space change all the rules. They offer us a boundless increasing-sum game, with wealth beyond our wildest Earth-bound dreams and opportunities for travel that boggle the mind.

15

CIVILIZATION
IN SPACE

"Good morning, Paula," sing-songed the instructional daemon with the smiling artificial face.

"Good morning, teacher," chanted Paula back.

"Did you have a good night's sleep, Paula?" Daemon of course already knew the answer.

"Yes, teacher."

"Did you have a nice breakfast?" Daemon's gallium arsenide intellect already knew every detail of the meal.

"Yes. I had muffins with blueberry preserves and cream cheese. And a glass of orange juice." And a glass of milk, added Daemon silently.

The ritual of human encounter satisfied, Daemon expertly, and in accordance with his programming, steered the conversation in the direction of the day's lesson plan. As always, Paula was given a chance to start well.

"Do you know where orange juice comes from, Paula?"

"Yes, teacher. It comes from oranges."

"Do you know where oranges come from?"

"Yes, teacher. They grow on orange trees."

That's right, Paula!" The artificial countenance again smiled warmly and nodded. "And where do orange trees come from?"

"They came from Earth long, long ago."

"Do you know where blueberries come from, Paula?"

"Yes, teacher. They grow on blueberry trees."

"Do you know where blueberry trees come from, Paula?"

"Yes, teacher. They came from Earth too."

"That is basically correct, Paula. Actually, they were engineered about five hundred years ago from blueberry bushes that came from Earth eight hundred years ago. Paula, do you know where all plants come from?"

Paula, torn between the self-confidence built by her earlier answers and her uncertainty on this particular point, hesitated. Daemon noted her hesitation and correlated it with the change in her cardiac and respiratory rates, pupil dilation, and the deflection of her eyes.

"I think all the plants came from Earth."

Daemon waited with an impassive, expectant simulated face.

"Except some of them, like blueberry trees, were engineered from Earth plants."

Daemon beamed.

Paula felt a little ripple of pleasure from making Daemon smile.

"Where do you come from, Paula?"

A flash of alarm lit Paula's brown eyes. She rehearsed a few possible answers while attempting to maintain an inexpressive face. Daemon noted her overt physiological changes and correlated them with her brain wave activity and pheromone emission.

"I came from my Mom and Dad," she offered tentatively.

"Yes, Paula, that's right. And where did your Mom and Dad come from?"

The safe pattern was clear. "They came from their Moms and Dads."

"And where did they come from?"

Paula had a momentary glimpse of infinite regression, though she had never heard of the concept. "I guess all the people there are came from Earth sometime."

"Good, Paula, that's right." Intellectual and emotional involvement indices were both above critical thresholds: time for the lesson to become more explicit. Time to integrate facts into the questions.

"There are ten quadrillion people, Paula. How many of them came from Earth? Earth holds only about ten billion people, and there are a million times that many in solar space."

Paula frowned. "I guess most people were born after they left Earth." She frowned harder. She didn't know anybody who had been born on Earth. "I mean, the people who left Earth had a lot of children."

"Good, Paula, that's exactly right. Now, what happened to the people left on Earth? Were there a lot of them?"

"Well, teacher, I know that there are still some people on Earth . . . " Paula trailed off, thinking furiously. "I guess almost everybody left, but a few stayed."

"No, Paula, it was the other way around."

Paula, unaccustomed to the word no, widened her eyes. Daemon observed and correlated seven other related changes.

"But, teacher, almost everybody is in space now. Doesn't that mean almost everybody left Earth?" Ah, thought Daemon, now she's asking the questions.

Daemon phrased the response to avoid negative words. "Only about one in every hundred people left Earth during any generation. Ninety-nine out of a hundred stayed there."

"So how many people were there on Earth when people started to leave? How many are still on Earth?"

"Eight hundred and sixty years ago, when people started moving away, Earth had a population of six billion people. It now has a population of seven billion."

"So Earth hasn't grown very much at all." Paula considered the problem. "So only one out of every hundred left. That's—let's see . . . " She frowned in concentration.

"That's less than a hundred million people. Actually, Paula, that's the most that ever left in any single generation. Overall, about a half billion people left Earth. Most of them left in the first three centuries."

"But in three centuries a lot more than six or ten billion people must have lived on Earth."

"Yes, Paula, about twenty-five billion people lived on Earth during that time. Only a half billion have ever left Earth since the beginning."

"Then that half billion who left have 10,000,000 billion descendants, and those who stayed sort of broke even."

"Yes, Paula, that's what happened. When people from Europe spread throughout Earth a thousand years ago, only about one person in a hundred left home. But they had as many descendants a few hundred years later as all those who stayed behind in Europe. The same thing *happened very early in the history of life on Earth, when the first plants and animals emerged from the oceans onto the empty land. The ones that moved from the ocean changed into a vast number of new species that came, in time, to dominate Earth."*

Paula nodded thoughtfully.

"Paula—do you know about the starship program?"

"Well, yes, a little. I know that about a million starships are going to be built to go to other stars."

"Now, Paula, how many people do you think will want to go?"

Paula thought deeply and answered correctly.

"Very good, Paula. That will be all for today."

Daemon logout.

—Civilization 101

Who will go into space? It is not always easy to leave, especially for those with the greatest need to leave. The worst tyrannies want stability at any price. They violently oppose emigration because it fosters the notion that there is a way out; it raises hopes. And it is rising expectations, according to Machiavelli, that cause revolutions. Subversives, idealists, inventors, artists, scientists, and mystics alike threaten the very fabric of tyranny: people of this ilk have a nasty way of asking questions such as, "Why do we have to do this? Why not that?" Forcing such people to conform is often much more difficult and expensive than letting them go.

Space will at first be largely a haven for refugees. The refugee state is usually a melting pot. New Amsterdam in the late 1600s was an amalgam of Huguenots from France, Walloons from the Low Countries, English Dissenters, Puritans, and Sephardic Jews from Iberia. They tolerated each other beautifully and got along well with the American Indians. But, as imperfect humans, they had a hard time forgetting how they had been driven from their homes in Europe and forced to emigrate: they were all there because they were escapees from Roman Catholic or Anglican persecution. Their common desire was to have religious freedom. Their tolerance faltered only in regard to their former persecutors. They found it hard to forgive their ancient oppressors and their government-established churches. If there were two things they all agreed on, it was the importance of freedom of religious expression and the inadmissibility of a state church. The modern attempt to put the atheistic concept of "freedom from religion" in their mouths would have been greeted with horror and dismay. It was the search for freedom of religion that brought most of our ancestors here, and it will be the search for freedom from religious, political, and ethnic persecution that will send the first colonists forth into space. Thereafter, anyone who wants to will be able to go.

But how many people will actually want to go into space? And how will they get off Earth? It is obvious from human history that not everyone will want to go. Suppose that only 1 percent of the human race chooses to depart in any one generation. As the human population grows from its near-future level of ten billion people, doubling every generation (thirty years), it can increase by a factor of ten every century. A population growth of a factor of one million would require only six centuries. Population pressure throughout the inner solar system could be a reality by A.D. 2600—unless, of course, we turn inward and stay on Earth. Then the classical limits to growth apply, and civilization collapses to subsistence agriculture by about 2030.

One hears glib assertions among some space fanatics that, if human meddling ruins Earth, we can always go somewhere else. This is the height of irresponsibility. If we ruin Earth and exhaust its energy resources, we will almost certainly have destroyed our means of escape along with everything else. The desire to evade responsibility for one's actions is a behavioral fossil, a relic of the days when nomadic bands of sloppy prosimians fouled their camp and then moved on to the next green valley to repeat their folly. It is not worthy of a human adult. Let us go for positive reasons or not at all.

But one conclusion seems very securely based on observed human behavior. No matter what happens and no matter how gruesome the prospects, some people will want to stay. No matter what ingenious device is invented, some people will refuse to use it. When electric lights were installed in the White House in 1889, neither President William Henry Harrison nor his wife could bring themselves to touch the switches. They left it to a janitor to turn on the lights each night and turn them off when he arrived at work the next morning. Many people would rather die than experience change, and many will get their wish. Thomas Edison, perhaps with the Harrisons in mind, commented: "Human inertia is the problem. Something in man makes him resist change."

But, for those who want to go, there can be an easy road off Earth. The first, and biggest, step is getting from the surface of Earth into orbit. In chapter 6 we discussed the prospects for dramatic reductions of the cost of launching payloads in terms of free competition, engineering innovation, hydrogen-oxygen propulsion, and airline-style operation. These factors seem capable of lowering launch costs to about $500 per kilogram. Once in space, the argument went, we have the option of refueling our vehicle with propellants brought downhill to Earth from the nearby asteroids or the Moon, rather than uphill out of Earth's deep gravity well. But, as we saw in chapter 1, even so cheap a launch is still considerably more expensive than the cost of the electrical energy needed to launch the payload. If we could run an "electric train" that is plugged into our commercial power grid from Earth into orbit, at the present cost of electricity (5 to 20 cents per kilowatt-hour), launch costs would fall from $500 per kilogram to about a dollar. Better yet, if we could use the power from a solar power satellite or deuterium/helium-3 fusion reactor at 1 cent per kilowatt-hour, the tab would be about 10 cents per kilogram. A 75-kilogram (165-pound) person could be launched for a price of $7.50! But is there any way to use electric power to launch people into space? Wouldn't an electrically propelled launch vehicle require a prohibitively long extension cord?

Wires are a very old-fashioned means of delivering energy, more characteristic of the nineteenth century than the twenty-first. The economy of the near future will have the option of delivering energy as a microwave or laser beam, transmitted through space without wires. Beaming megawatts of power to a rocket in flight will be a fairly simple exercise. Now suppose our launch vehicle has such a "virtual extension cord." What will it use for its exhaust gases? Two possibilities come to mind: air and water. Both have such a low market price and are so ubiquitous in the environment that they can be used in vast amounts without incurring either great expense or environmental degradation. Here we have all the advantages of a skyhook without the enormous cost of building it. In addition, we free ourselves from the fundamental limitation of a skyhook—the requirement that it operate exclusively in the two-dimensional equatorial plane of the planet to which it is anchored. (Highly eccentric earth orbits, or HEEOs, the ideal locations for refueling bases, are also very hard to reach from a skyhook.) But the laws of physics let us operate at very low cost in three dimensions. Why throw one of them away? And why should we pay ten thousand dollars a kilogram to put payloads into space, when it only takes pennies' worth of energy to do the job? *The idea that space travel is inherently enormously expensive is fraudulent.*

Once we are in space, whether in a classical low space station orbit (LEO) or in an elliptical orbit reaching out to the Moon (HEEO), we have the option of filling our tanks with water from asteroids, water from Earth, hydrogen-oxygen electrolyzed by solar power from water, or just liquid hydrogen. We therefore can refuel nuclear thermal rockets that use either water or hydrogen as the working fluid, solar thermal engines (with the same choice of fluids), or hydrogen-oxygen chemical rockets. These very diverse propulsion systems permit a wide range of possible missions, ranging from Moon landings to outerplanet probes. And in the long run, deuterium from Earth and helium-3 from the Moon and Uranus will be available in HEEO to fuel fusion-powered spacecraft for interstellar probes.

Once we have *any* access to HEEO, "beltstrapping" our resources by returning NEA water and structural materials can provide at least another factor-of-100 leverage on the amount of materials available to us. Piling this advantage on top of cheap launch from Earth makes our wildest dreams look pitifully unimaginative. The science fiction future in which rich "spacers," liberated by near-magical technology, flit about the solar system while miserable, ignorant, sickly, impoverished "groundhogs" live in squalor on a teeming, hideously polluted Earth is just a bad dream. The fault of this future is not that it contains too much technology, but that it contains far too little. But remember that such a dystopian future is still within our power to create.

It would be quite incorrect to say that my vision of the future is optimistic—I do not know what path will be chosen. I am well aware that humans are capable of exercising free agency. The future can (like the past) hold both very good and very poor choices. I am content to point out one or two of the many paths that do not lead to the end of civilization. I am quite certain that I cannot imagine all the promising options—or all the hideous ones. But the door stands open to a golden age greater than any we have ever seen.

.∴.✳.

And what will we do when the solar system is full of hundreds of quadrillions of people? Can they in turn move on, or must they, like Earth, stagnate in a steady state? Two elements are at work here: how many want to go and how many can be transported with the available resources. If we suppose that the solar system's population levels off at about 10,000,000 billion, and 1 percent of them wish to emigrate in any given generation, than the departure rate would be 100,000 billion per thirty-year interval, or 3×10^{12} people per year. The resources required to support this emigration rate would consist of their own fair share of the total resources available, plus enough helium-3 to transport them to another stellar system. The solar system, its population limited by the need to recycle all materials and run on solar power, can grow beyond 10,000,000 billion only by exploiting new resources beyond those present in the asteroid belt, such as the Trojan asteroids and the small outer satellites of the giant planets. Science fiction contains many stories about mining comets out in the Oort comet cloud, thousands of astronomical units from the Sun, and even about making food from the organic materials in them to feed the "teeming billions of Earth." But the energy cost and time requirements for such trips essentially assures that they will be one-way journeys. We must remember that such speculations were written by people ignorant of the vast wealth of volatiles in the outer belt, the Trojan asteroids, and the small outer satellites of Jupiter. Minimum-energy trip times from Earth to Jupiter's system are only three years, and somewhat larger energy expenditures could significantly reduce these trip times. There are already several asteroids known in the near-Earth population that pass very close to Earth and have aphelion distances close to Jupiter's orbit. These bodies are likely to be extinct comet nuclei, rich in ice and other volatiles. And of course, access to the Jovian system from the belt is much easier and quicker than from Earth.

It is conceivable that an impending solar system—wide disaster might encourage almost everyone to pack up their bags and leave. There are examples in history in which the majority of the population of a country left: the Irish in the mid-1800s during the potato famine, the apparent Dineh (Navajo) migration in the sixteenth century from the Pacific Northwest to Arizona, and the seventeenth century Basuto migration into South Africa are well known. A threatened calamity on the scale of a major change in the luminosity or stability of the Sun would encourage almost everyone to leave, but solar physicists project several billion more years of stability before the Sun begins to age off the main sequence, cooling and expanding to become a red giant. A nearby supernova explosion that imperils the entire solar system would seem an adequate threat, but there is neither any evidence of a nearby pre-supernova star, nor any hope of outrunning the blast wave from such an explosion, should one occur.

In chapter 12, I described six metaphors to give us some idea of the vastness of the resources of the asteroid belt. To these, we can now add a seventh. *Perspective 7:* The Sun is old and we need to go somewhere else before it bloats and cooks us. We have enough asteroidal iron to build a very large spaceship. Imagine a cylindrical spacecraft with a diameter of 1 kilometer (3,300 feet). We have enough steel to make that spaceship four billion kilometers (2.4 billion miles) long— long enough to reach from the Sun to beyond Uranus! With a ship that long, there is no need to travel—you are already there! If spaceship lengths in excess of, say, 1,000 kilometers seem impractical or immodest, then we could simply build four million 1,000-kilometer-long spaceships. Each one would then be able to hold a mere 2.5 billion people. Or how about a fleet of 4 billion kilometer-sized ships? *That's enough to send 2.5 million people to each Sun-like star in the Galaxy.* Each such ship is self-sufficient in materials (through recycling) and energy (through fusion). It is, in effect, a small traveling world.

The option of departure will open up as soon as the deuterium/helium-3 fusion economy is established. That could in principle be as early as A.D. 2030, but that would require intelligent planning. Fortunately, the exact date scarcely matters: technology advances so rapidly that any impossible dream becomes conceivable, then possible, then routine, and then obsolete over the course of a few decades. The key technology to permit interstellar migration, which seems at this point to be the fusion rocket engine, I will tentatively assign (with all due conservatism) to thirty to sixty years from now.

There have been many speculations about advanced means of traveling between the stars. One of these is matter-antimatter propulsion. Another is electrical propulsion in the form of ion engines, plasma jets, or the like. A third is riding an intense beam of laser light on an enormous, diaphanous sail of metal foil. We shall ignore such options, not because they are physically impossible (they are not), but because we do not need them to make my point. If history shows my extrapolations to be too conservative, I will be neither surprised nor disappointed. Quite the contrary. There are many ways to empty a solar system.

A vast human population can be packaged for export in several possible ways. First, the simplest and most expensive, is to send them off as stable, living, closed ecological systems, running on fusion power. These would be real, living, breathing humans with lots of elbow room and an unlimited opportunity to view new scenery. Science fiction writers term these traveling worldlets "generation ships." At 100 kilometers per second, such a ship could reach a nearby star in 5,000 years. At 1,000 kilometers per second (1 astronomical unit per day) the trip would be reduced to under 500 years. (This is the same time scale as the terrestrial age of exploration: 500 years ago Columbus was on his third voyage to the New World.)

Second, there is the science fiction device of sending out stacks of boxes of people in cold sleep (referred to by some waggish SF authors as "corpsicles"). The sleepers would be awakened when the ship arrives at its destination. A small, rotating crew would be awake at any time to oversee the maintenance and navigation of the ship and the welfare of its occupants. This second option, which may become technically feasible within a few decades, could be based on any of several variants of the "cold-sleep" idea. Many levels of "sleep" (and of cold) appear conceivable, ranging from artificial hibernation at a body temperature barely above the freezing point of body fluids to deep cryogenic storage at temperatures approximating absolute zero.

A third option is to ship vast numbers of future people in the form of frozen eggs and sperm. Large "classes" of children would be raised, either by a cold-sleep crew of adult human teachers or by educational robotic "parent surrogates," upon arrival at the destination. This is certainly the most efficient of the options.

The most demanding case, requiring the most mass, volume, and power, is the first. Consider an "interstellar ark" made of asteroidal metals and powered by helium-3 and deuterium. This ark, a sphere with a diameter of 10 kilometers, has a volume of 100 cubic kilometers and a mass of 2.5×10^{15} grams (2.5 billion tons). It houses a billion people. To accelerate this ark to 1,000 kilometers per second uses 10 million tons of helium-3. There is enough mass in the asteroid belt to build a billion of these arks. The total amount of helium-3 required to send them all off on journeys to other stars is 10 quadrillion tons (10^{22} grams) of helium-3. We saw in chapter 13 that the amount of helium-3 in Uranus and Neptune is 500 times as large as this.

It is amusing, and often instructive, to ask what we could accomplish with present-day and near-future technologies. We should remind ourselves that our vision of the next few centuries has so far been based on conservative technological assumptions. We have generally assumed technologies that are within or near current capabilities. Solar power satellites, for example, could have been built with 1964 technology, although they would not then have been economically competitive. Almost every industrial process we have described for processing nonterrestrial materials has been experimentally investigated in the laboratory. Fusion power generation has made enormous progress since 1960. The challenge with fusion is to increase the plasma density and its containment time until the rate of energy production, which is proportional to both these quantities, greatly exceeds the energy consumption needed to run the machine. Using the product of the density and the containment time as a yardstick, fusion technology has advanced by about a factor of 10^{12} in the past thirty-five years. The ITER project (see chapter 8) should advance us another factor of ten, putting the design of a commercial fusion reactor within reach. The fusion-engine spacecraft is a more ambitious and uncertain extrapolation. Perhaps we would be wise to separate the feasibility of the fusion power *concept* (high) from the feasibility of *current fusion reactor designs* (quite uncertain).

Future civilizations will not, of course, be constrained to rely only on technologies that are already in hand or clearly feasible within the next few years. The events of A.D. 2060, for example, will take place with the advantage of more than sixty years of technological progress beyond that allowed in the conservative extrapolations just cited. Sixty years of modern scientific and technological innovation cannot be safely neglected. Each generation brings onto the scene a host of powerful new tools, multipliers of human effort, that increase our productivity and wealth. Any specific job becomes easier with time and costs less to do as the physical and mental powers of machines increase. Compare, for example, the state of aviation, rocketry, space exploration, electronics, transportation, communications, and computer technology in 1910 to that in 1970. Although it is impossible to foresee all the new capabilities that may arrive over the span of the *next* sixty years, it is still possible to foresee some of the directions in which current scientific and engineering will progress.

In this spirit of more liberal technological assumptions, we can predict a number of major advances that will occur within this two-generation horizon. One of these areas, advanced spaceflight technology, has already been summarized in sufficient detail.

We can also confidently expect computer technology to continue to advance at a great rate. We may expect the processing speed of computers, their information-storage capacity, and their random access memory capabilities, to increase by about a factor of one thousand every decade. As a rule, each generation of computers is somewhat less expensive than its predecessor. Within thirty years, present magnetic storage devices and electronic circuitry will give way to optical storage and photonic computing at the speed of light. The logical course for the following generation is miniaturization of photonic computing. By then, we should expect extremely compact personal computers with very low power consumption and performances in excess of the largest supercomputers of today. Highly efficient and intelligent computers greatly increase the amount of work that can be done by each person, and therefore lower the cost of large projects. When most grunt work is done by semi-intelligent, semi-autonomous computers and robots, human productivity (and human wealth) will be multiplied tremendously.

The capabilities of present computers are already so great that it would take decades for writers of software to tap all their potential. Extremely complex software programs, called expert systems, are now available for a number of specialized purposes. We can expect enormous growth in the power and scope of expert systems as computer hardware continues to evolve. Imagine a computer designed to serve as a consulting physician, equipped not with what was taught in medical school twenty years ago but with complete access, through computerized translation, to the entire medical literature in every language. Every medical facility, whether in Tokyo or a remote mining town on the asteroid Ceres, would have immediate access to every diagnostic means, laboratory technique, surgical procedure, and billing protocol known.

Highly capable expert systems can easily be linked via computer networks to create a computer system that has access to all human knowledge. The individual can link into this network and access the contents of the world's libraries and archives. The normal form of publication will be entry into some computer-readable online storage medium. This is scarcely a bold projection, since the hardware and software to do this are already in production or in development.

Communication between complex programs and human users can be improved by programming the computer to emulate the behavior, speech patterns, and reasoning of humans. Such a program, called an artificial intelligence, is in effect a computer simulation of a human being.

The bottleneck in the information system of the future is the human–computer interface. Expert systems capable of understanding human speech and of reading written text aloud already exist and are improving rapidly. The need for keyboards to enter information is now approaching its high-water mark. Verbal interaction with computers, beginning with but certainly not limited to dictation of text, will soon be the normal and preferred means of input. The eye can serve as a kind of cursor to specify points on a display screen, monitored by special sensors that determine the eye's point of focus. Three-dimensional graphical displays, both holographic and mechanical (using laser illumination of a spinning helical target) already exist. Both input and output of data will be vastly improved, narrowing the gap between humans and machines.

Because of the extreme portability of future personal computers, minimization of the bulk of the interface equipment will be extremely important. It is not yet possible, except in some very limited applications, to design and build a direct interface between a computer and a human brain. Nonetheless, experiments with computer simulation of vision for the blind and computer-generated hearing for the deaf have shown the promise of such direct brain–computer links. Expected advances in processing power make such links ever more feasible and likely. Implants of computer hardware, such as artificial eyes and organs that sense at wavelengths other than those directly perceivable by biological senses, are near-term possibilities.

Artificial intelligence can feasibly be combined with a mobile platform to make a device that is the functional equivalent of a simple living organism. Following the lead of the Czech science fiction writer Karel Capek, such devices are called robots, from the Slavic root *robot-*, which means "worker." Many simple robots have already been built for specialized tasks, but none even remotely approach the adaptability and level of intelligence of a human being. The average family can look forward to having routine house and yard work performed by robotic devices within a few years. Robots are of special interest in hazardous environments, in which a human would be at risk but a robustly designed robot would be safe and functional.

The long-term evolution of robotics will lead inevitably to the design of robots that emulate not just a few human traits but full human capabilities. It would then be tempting to package such a robot in an imitation human body. Such robots, which have the approximate functionality and appearance of humans, are termed androids. From the writer Isaac Asimov's android detective R. Daneel Olivaw to present-day TV characters, androids raise countless fascinating questions. For example, supposing that an android can successfully pass as a human being in human company. Never mind if it seems to be an odd human or a stupid human—is it credible as a human? If so, what grounds do we have for distinguishing between the android and a human of equal intelligence? Would such an android deserve the protection of law? Citizenship? The right to vote? The right to run for public office?

Where do we draw the line? And what about vastly superior computer minds that happen not to reside in humanoid bodies? Readers of Asimov quickly observe that he presents androids as parables of ethnic minorities, and he parallels the search for acceptance and experience of alienation felt by both.

We are also on the threshold of new technologies for communicating with each other and with remote computers. The last few years have seen a tremendous proliferation in facsimile transmission of documents (a technology first used commercially during World War II) and electronic mail. Voice mail is in widespread use. Cellular telephone usage and electronic mail are undergoing explosive growth. Several industrial consortia have formed to compete for a vast new market for global satellite-based cellular phone service. Since the technologies for transmission of text documents, graphs, and voice mail are already well adapted to telephone links, we are only a few years away from worldwide, twenty-four-hour access to every other person on Earth. Satellite radio and TV links already span the world, and optical-fiber cables serve vast and growing populations. Satellite links with extremely high data rates using lasers instead of radio waves are difficult on Earth because of the intervention of clouds, but links from geosynchronous orbit outward are free to evolve in this direction. In the foreseeable future, communications will be so good that many of the incentives to travel will be reduced. In effect, communications and transportation are already beginning to compete with each other.

One of the most interesting capabilities of high-speed computers is to build up a three-dimensional simulation of reality, which can be viewed in a special room or through goggles or eventually by direct neural linkage of a computer to the optic nerve. Such computer-simulated spaces are called virtual reality. Using a high-data-rate communications link and a computer simulation of, for example, a Paris street scene, permits the illusion of being actually present at another location (or several, for that matter). A salesman may achieve telepresence at places he has never visited in the flesh. A worker in orbit about Venus could link to sensors on a rover touring the hellish inferno of the lowland plains, where the ground glows a dull red and the air is hotter than a self-cleaning oven. A geologist inside a spacious, comfortable lunar-base module could drive a breadbox-sized prospector buggy around on the lunar surface, "seeing" the surface with far more sensors than any biological organism ever had. A meteorologist could "ride" a hot-air balloon through Jupiter's Great Red Spot from the safety of the surface of Callisto, directly sensing every nuance of the atmospheric motions, heat flow, and exotic chemistry.

The ability to operate in such environments permits us not only to explore other worlds, but also to operate on them. A telepresence engineer in orbit high above the surface of Mars could run, maintain, and repair an automated liquid oxygen plant on the surface, accumulating propellant and air for use by an expedition that has not yet arrived. Or she could remotely operate equipment that sprinkles dark Martian dust over the polar ice caps, warming them and releasing large amounts of carbon dioxide and water vapor into the atmosphere to build up the atmospheric pressure. She could operate a plant that extracts chlorine and fluorine from crustal minerals, reacting them with atmospheric gases to make CFCs to bolster the greenhouse effect. Or she could supervise the work of billions of lithophages as they consume and package for export an asteroid that threatens to collide with Earth in a few hundred years. Thus, teleoperation paves the way for planetary engineering, the systematic alteration of planetary environments to make them more hospitable to humans. Machine intelligence makes possible operations on a scale we could not otherwise manage.

One of the implications of ever-smaller, ever-more-powerful computers is the ability to build autonomous machines at the cellular size level. This is the emerging field of nanotechnology. The pioneering research of M.I.T. nanotechnologist Eric Drexler has shown in considerable detail the constraints on—and the capabilities of—these tiny "engines of creation." They contain the potential for building custom materials, both inorganic and organic, an atom at a time. Nanomachines could be injected into the bloodstream to seek out toxins or infectious organisms, or even tailored to correct genetic defects directly in the DNA. Or they could extract platinum metals from asteroidal iron, excreting bricks of stainless steel to be put to other uses in space. Or, ultimately, such machines may be employed in making copies of themselves.

The feasibility of designing and building machines capable of replicating themselves was first studied in the 1940s by the brilliant mathematical physicist John von Neumann, of the Institute for Advanced Study in Princeton, New Jersey. Given a supply of components, it is demonstrably possible to design a machine that can assemble exact copies of itself from those components. The simpler and more numerous the components, the more demanding the assembly task becomes. In the extreme case of replication from natural raw materials, very elaborate processing and fabrication are required, especially for the most complex part of the machine (its electronic brain). Truly self-replicating machines that can ingest natural raw materials could potentially be the most dangerous and destructive vermin in the universe. For this reason, some external control over the replication and function of nanomachines must be exercised. Two simple precautions let us remove the potential for cancerous proliferation of these machines. First, we should not give them the ability to replicate their controller chips ("brains"), but instead manufacture them separately and supply them as needed. Second, we should design them to run only off of external power sources that are not available in nature, such as beamed microwave power. We could then shut them down by the simple expedient of "pulling the plug."

Mapping of the human genome and research enabled by having a genome map make several types of intervention by subcellular machines possible. One function of such bioengineering would be genetic correction, fixing the results of damaging mutations. How much better it would be to prevent such damage, not merely correct it, often imperfectly, after it happens! The charge that fixing genetic damage is "playing God" ignores the very substantial contribution that surgery has already made to the correction of genetic defects after birth, after the damage has already been done. Since the benefits are manifestly obvious to millions of relatives and friends of those who have undergone corrective surgery for genetic defects, few people are curmudgeonly enough to level the same charge against that practice. Opposition to corrective genetic manipulation seems to be yet another case in which people would rather die than change; however, in this case, the message is that they would rather see someone else die than be changed. Whereas they are certainly free to make that decision for themselves, the desire to exercise the power of compulsion against victims of genetic damage to prevent their permanent cure (not just cosmetic correction!) is obviously unrighteous.

A second aspect of the ability to change genetic information concerns the introduction of new or modified genes, an area that may properly be termed genetic manipulation. Science fiction is replete with cosmetic and vanity uses of genetic engineering. These may have entertainment value, but no long-term importance. The real significance of genetic engineering is that such techniques can lead to directed human evolution, including modified design of humans to help adapt them to alien environments. We may modify planetary environments greatly through planetary engineering, but it may often prove more feasible to "meet the planet halfway" by developing human genetic strains better suited to that environment. Such adaptation is a natural part of human evolution on Earth, proceeding via the inefficient and slow process of natural selection. On Earth, clans that move into harsh Arctic regions are genetically pruned by their environment: tall, thin bodies are harder to insulate and keep warm in deep subzero temperatures; long fingers, noses, and ears are prone to frostbite and subsequent infection and loss. Natural selection in the Arctic brutally, and at the cost of countless personal tragedies, selects in favor of stocky and compact bodies, flat noses, small ears, and stubby fingers. Nature favors Eskimos in the polar climate. Clans that populate extremely hot, sunny areas are also genetically shaped by factors such as skin cancer and hyperthermia. Natural selection in the deserts favors highly pigmented skin and tall, lean body types. More generally, in ancient times our ancestors, who were probably residents of warm climates and highly pigmented, dispersed to more diverse climes where the Sun was less brutal and where mutations leading to other, lighter skin colors were no longer significantly disadvantageous. Of course, all human "races" are members of a single species, capable of fertile interbreeding, that have diverged from their common ancestors to only a limited degree. But future genetic drift can be directed with far less human tragedy, to result in a far broader range of progeny.

In the great diaspora to come, deviations of different emigrant groups from each other will occur due to initial selection from different gene-pool mixes, intentional genetic manipulation to fit different environments, natural selection in vast numbers of very different solar and planetary environments, and random point mutations (each of which of course occurs in only one "branch" of the human genetic tree). The different shoots from the stem of Earth will be isolated from genetic contact with each other for tens of thousands of years at a time, far longer and with incomparably better genetic isolation than any gene-pool separation ever achieved on Earth. Especially favorable mutations may be treasured as "corporate secrets" of a particular branch of humanity, or the DNA coding of that mutation may be bartered for other specialized knowledge via closed-beam laser communication links with other exploration and colonization parties, or it may be freely disseminated as part of a "galaxywide web" of human communication.

The inevitable result of prolonged genetic isolation and adaptation to very different environments will be that one after another of the offshoots from humankind will become incapable of breeding fertilely with each other. Humankind will undergo speciation, giving rise to billions of new species descended from twenty-first-century humans. *Biotechnology, including the ability to build subcellular machines, is one of the most protean contributors to our movement into space.*

If biological speciation is certain, then cultural speciation is even more so. Language drift is fast enough so that few languages are mutually comprehensible after a thousand years of divergent evolution. Genetic connections between languages that diverged several thousand years ago are, if terrestrial experience is a fair indicator, barely even discernible to the trained historical linguist. But all these human societies will start off in contact with each other and in constant communication. They may hold on to a common formal communications language long after it fades out of daily use, in the fashion of Latin in medieval Europe or Sanskrit in India. Or, of course, being humans, many of them will prefer to use their own native language or some pidgin to distance themselves from the rest of humanity. It is not too difficult to imagine two colonization ships following similar paths, speeding along within hailing range of each other, for thousands of years—one insisting on speaking Slovenian and the other adamantly broadcasting only in Ossetian.

Technological innovations will occur randomly throughout the expanding human swarm. Some clans will dart ahead in computational techniques, or neo-representational poetry, or show-rabbit breeding, or fusion-power-plant engineering, or human genetic manipulation, or kinetic light sculpture, or food production, or hyper-Boolean music, or the construction of centrifugal bumble-puppies and leave the rest of the universe far behind. Cultural obsessions and transient fads, such as tulip growing in Turkish and Dutch history, Hula Hoops in America, or Communism in Russia, are utterly unpredictable but may have long-term cultural repercussions. It may be that our human branches will simply lose interest in each other because of their cultural divergence. The reader who wishes a prescient glimpse of a very distant, very alien human future would still be well advised to read Olaf Stapledon's 1930 book, *Last and First Men.*

Ultimately, as we move out into the galaxy, we shall encounter a vast range of new stellar and planetary environments. With the tools at our disposal we shall shape planets in the image of Earth and remold humankind and other species in the context of alien planets. Billions of races, trillions of species, and numberless souls will flow forth from our hands.

Let us recapitulate what we have already found. *Shortage of resources is not a fact; it is an illusion born of ignorance.* Scientifically and technically feasible improvements in launch vehicles will make departure from Earth easy and inexpensive. Once we have a foothold in space, the mass of the asteroid belt will be at our disposal, permitting us to provide for the material needs of a million times as many people as Earth can hold. Solar power can provide all the energy needs of this vast civilization (10,000,000 billion people) from now until the Sun expires. Using less than 1 percent of the helium-3 energy resources of Uranus and Neptune for fusion propulsion, we could send a billion interstellar arks, each containing a billion people, to other stars. There are about a billion Sun-like stars in our galaxy. *We have the resources to colonize the entire Milky Way.* But there is no need to wait several billion years before beginning the cosmic diaspora; we could begin within a few decades. But we could dedicate ourselves *today* to finding the nearby asteroids that threaten us, the sword of Damocles that hangs above our heads—and beating that sword into plowshares to serve the future of humanity.

As long as the human population remains as pitifully small as it is today, we shall be severely limited in what we can accomplish. Human intelligence is the key to the future: human beings are not, as some would have it, a form of pollution. Having only one Einstein, one Hokusai, one Mozart, one daVinci, one Shankara, one Poulenc, one Arthur Ashe, and one Bill Gates is not enough. We need—and can have—a million times as many. We need intelligence, wisdom, compassion, and excellence. These godlike traits are manifested in the physical universe only by life and in the biological universe only by intelligent life. Life is not a cancer of matter; it is matter's transcendence of itself. We would be quite wrong to conclude that the asteroid belt or solar power or helium-3 in the Jovian planets is the ultimate resource. *Intelligent life, once liberated by the resources of space, is the greatest resource in the solar system.*

The material and energy resources of the solar system allow humankind an infinite future: we can not only break the surly bonds of Earth but break free of the Sun and escape its fate. The fulfillment of time and space is matter; the fulfillment of matter is life; the highest fulfillment of life is unbounded intelligence and compassion.

Till when the ties loosen,
All but the ties eternal, Time and Space,
Nor darkness, gravitation, sense, nor any bounds bounding us.

Then we burst forth, we float,
In Time and Space O soul, prepared for them,
Equal, equipt at last (O joy! O fruit of all!) them to fulfil O soul.
                                                      —Walt Whitman

There is no way back into the past.
The choice is the Universe—or nothing.
                                                      —H. G. Wells

AFTERWORD

Several developments relevant to the subject of this book have occurred since its publication in the fall of 1996. These include reports of evidence for possible ancient life on Mars, several spacecraft missions either launched or planned to explore Mars, the discovery of a new class of massive icy bodies far from the Sun, reports of exciting new experiments on unconventional propulsion techniques for space launches, and news of plans for a privately funded prospecting mission to one or more Near-Earth Asteroids.

In September 1996, a team of researchers headed by Dr. David McKay of Johnson Space Center reported that careful examination of a Martian (SNC) meteorite, AlH 84001, revealed that it contained traces of complex organic molecules called PAHs (for polycyclic aromatic hydrocarbons). Further, the meteorite contained tiny particles that resembled fossil bacteria. Several lines of evidence were developed that were compatible with the idea that this meteorite, long before it was blasted off of Mars by an impact event, may have contained living cells. Unfortunately the evidence, however interesting, is in no way diagnostic of the former presence of life. Non-biological explanations also exist for all the phenomena reported; for example, PAHs can be made by a variety of non-biological processes and are present in the Antarctic ice from which the meteorite was recovered. A vigorous, and sometimes vitriolic, debate is currently raging about whether the fossil-like objects originated at low temperatures (compatible with a biological origin) or at high temperatures (suggestive of an inorganic mineral reaction).

It is important to realize that the actual experiments reported by the NASA-JSC team are not under attack: the team involved has a lengthy history of solid, reproducible work on lunar and meteorite materials. Rather, it is the uniqueness of their interpretation of these experiments that is at issue. The dilemma presented by this debate is quite trying for many planetary scientists, since it is widely accepted that the presence of primitive life forms on the ancient surface of Mars is plausible. But so far-reaching is the conclusion that merely suggestive results are not enough. As of this writing (July 1997) no convincing proof of biological origin has yet been presented.

Mars was again in the news in November and December. On November 6 the *Mars Global Surveyor* was launched from Cape Canaveral aboard an Atlas II booster. Assembled largely out of spare parts from the ill-fated *Mars Observer* mission of 1992–93, MGS is due to arrive at Mars for an orbital mapping mission in July 1997. The Russian *Mars '96* spacecraft, carrying an ambitious assortment of components including an orbiter, multiple landers, and surface penetrator probes, was launched from the Baikonur Cosmodrome in Kazakhstan on November 16. It was unfortunately lost due to the failure of its Proton launch vehicle to achieve orbit. The vehicle burned up in the atmosphere near the coastline of South America. Finally, the *Pathfinder* spacecraft was launched from Cape Canaveral on an Atlas II booster on December 2. This spacecraft includes an orbiter, a lander, and a compact surface rover named *Sojourner*. American plans for Mars include at least eight further small missions in the next decade, of which the next will be the *Mars Surveyor '98* mission, incorporating both an orbiter and a lander. Japan has announced its intention to launch the *Planet B* spacecraft to orbit Mars in the same launch window. The ambitious Russian *Mars '98* mission, originally planned to carry a large *Marsokhod* rover and a French-built balloon-borne instrument package, has been delayed by fiscal stringencies in the Russian space program. It has been tentatively rescheduled for launching in 2001, probably with one or the other of the two subsidiary payloads left off. The first experimental attempts to manufacture propellants out of the Martian atmosphere are expected in 2001. By 2005 NASA should be ready to land the *MISR* spacecraft and its internal chemical processing plant on Mars, manufacture enough propellant for the return trip to Earth, then blast off (in 2007) and return samples of the surface and atmosphere of Mars to Earth. The lengthy drought in Mars-oriented space exploration, extending from 1975 to 1992, has clearly ended with a blizzard of activity.

In June 1997, Jane Luu of the Harvard-Smithsonian Center for Astrophysics reported the discovery of a massive icy body, named 1996 TL66, in orbit about the Sun at a distance of about 100 Astronomical Units, about three times the distance of Pluto. This body, about 500 kilometers in diameter, has about 3 percent of the mass of the Moon and pursues an orbit that crosses the orbits of several of the outer planets. It appears that 1996 TL66 is but one of a vast swarm of icy bodies forming part of the "Kuiper Belt," a family whose total mass and resource content may surpass that of the asteroid belt by a factor of 1000 to 10,000.

The past few months have also seen an exciting series of tests of model launch vehicles propelled by external sources of power, such as beamed microwave power and ground-based lasers. Professor Leik Myrabo of Rensselaer Polytechnic Institute has carried out a series of tethered flight tests of small vehicle models at the MIRACL mid-infrared laser facility in New Mexico, demonstrating rapid acceleration of a fuel-less, engine-less vehicle under the impetus of an external pulsed high-power laser.

On July 4, 1997, *Pathfinder* and *Sojourner* landed successfully on Mars.

Finally, at the biennial Space Manufacturing Conference run by the Space Studies Institute of Princeton, space-development entrepreneur James Benson announced the existence of a consortium of scientists, engineers, and financiers dedicated to the construction and launch of an unmanned asteroid prospecting mission. Current plans suggest a launch as early as 2001 to send the spacecraft to the surface of a selected near-Earth asteroid for the purpose of filing a mining claim on the body. Since several NEAs have Earth-surface cash values in excess of $1000 billion, this mission constitutes a truly unique investment opportunity.

SUGGESTED READING

Among the most important of the early visionary writings on the resources of space was Konstantin E. Tsiolkovskii's *Exploration of Cosmic Space by Means of Rocket Propulsion*, published in Russian in
1903. The first expression of such ideas in a western European language was J. D. Bernal's essay, *The World, the Flesh, and the Devil: An
Enquiry into the Future of the Three Enemies of the Rational Soul*,
published by K. Paul, Trench, Trubner and Co., Ltd. of London in
1929. A fictional amplification and extension of these ideas appeared
soon thereafter in Olaf Stapledon's novel *Last and First Men*, published by Penguin at London in 1930.

Although Tsiolkovskii and Goddard (in an unpublished essay) both
emphasized the importance of asteroids as sources of industrial raw
materials, the first proposal to exploit near-Earth asteroids was that of
Dandridge M. Cole and Donald W. Cox in their 1964 book *Islands
in Space: The Challenge of the Planetoids*, from Chilton Books in
Philadelphia.

The idea of large-scale industrial operations in space (though with
no attention to extraterrestrial resources) was developed in a 1975
book by G. Harry Stine, *The Third Industrial Revolution*, published in
New York by Ace Books. Harry tells me that a new edition is in the
works. A broad vision of the use of the energy and material resources
of nearby space, leading to large-scale space colonization, appeared
in Gerard K. O'Neill's 1976 book *The High Frontier*, from Morrow in
New York. Gerry O'Neill's concept of building solar power satellites
was at the economic basis of this scheme. His vision was extended

and updated in his 1981 book, *2081: A Hopeful View of the Future,* published by Simon and Schuster in New York. In large part motivated by the public interest surrounding O'Neill's books, NASA produced a 1977 study, edited by Richard D. Johnson and Charles Holbrow, titled *Space Settlements: A Design Study* (NASA SP-413). O'Neill's sometimes collaborator, the former astronaut Brian O'Leary, expanded on space resource use in his 1981 book *The Fertile Stars,* from Everest House in New York.

An analysis of the relationship of the Soviet and American space programs to space resource use appeared in the 1987 book *Space Resources: Breaking the Bonds of Earth,* by John S. Lewis and Ruth A. Lewis, from Columbia University Press.

Two important studies of future space policies, motivated in part by the *Challenger* disaster and in part by the emergence of new ideas such as space resource use, appeared in the 1986 report of the National Commission on Space, *Pioneering the Space Frontier,* published by Bantam Books. The Synthesis Group report in 1991, titled *America at the Threshold,* was similarly visionary; however, the slowly ebbing NASA budget effectively prevented these progressive concepts from having much impact on real flight operations.

Low-cost launch prospects are discussed in G. Harry Stine's 1996 book, *Halfway to Anywhere,* from M. Evans, New York.

There are two major recent technical volumes that explore the availability, extraction, processing, use, and economics of space resource use. The first, a report on a workshop held at the California Space Institute in 1984, was published in 1992(!) in a boxed set of softbound books under the title *Space Resources,* edited by Mary Fae McKay, David S. McKay, and Michael B. Duke (NASA SP-509). The other volume, based on a 1991 conference at the University of Arizona, and edited by John S. Lewis, Mildred S. Matthews, and Mary L. Guerrieri, is *Resources of Near-Earth Space,* published by the University of Arizona Press of Tucson in 1993.

Readers interested in an introduction to the astonishing possibilities of nanotechnology are encouraged to read K. Eric Drexler's *Engines of Creation,* a 1986 publication from Anchor Press, New York.

GLOSSARY

Asteroid Belt   A region extending from about 2.2 to 3.3 AU from the Sun, within which most asteroids are found.

AU   Astronomical Unit; the mean distance of Earth from the Sun. About 150,000,000 kilometers or 93,000,000 miles.

C asteroid   Carbonaceous asteroid. Rich in organic matter, water-soluble salts, magnetite, and clay minerals, they dominate the outer asteroid belt.

CFC   Chlorofluorocarbon. Any of a family of compounds of carbon, chlorine, and fluorine, commonly used as refrigerants or as aerosol propellants.

Chondrite   Meteorite (asteroidal) materials about 4.55 billion years old that are "primitive" and have never experienced melting and density-dependent differentiation.

Delta V   Velocity change.

Deuterium   A stable heavy isotope of hydrogen, containing one neutron and one proton in its nucleus.

Electrolysis   Decomposition of a substance by electrical current.

ESA   The European Space Agency.

Eunuch   An ancient term in use for millennia; a synonym for *bureaucrat*.

GEO   Geosynchronous orbit. An orbit at an altitude of about 39,000 kilometers above Earth's surface, with an orbital period of exactly one Earth day.

HEEO   Highly eccentric Earth orbit. Any of a vast variety of eccentric orbits about Earth, usually with perigees of a few hundred to a few thousand kilometers and apogees of tens of thousands to hundreds of thousand of kilometers. HEEO orbital periods usually range from a day to two weeks.

Helium-3 A stable light isotope of helium with one neutron and two protons in its nucleus. Helium-3 is highly reactive at stellar interior temperatures (ten million degrees), and is a potential commercial fusion fuel. Also written ^{3}He.

Ilmenite The iron-titanium-oxide mineral $FeTiO_3$.

ITER International Thermonuclear Experimental Reactor. A commercial-prototype fusion power reactor under development for testing in the late 1990s.

KREEP A rare but interesting type of lunar rock, named for its high content of potassium (K), rare earth elements (REE), and phosphorus (P).

LEO Low Earth orbit. A typical low-altitude (300- to 600-kilometer) near-circular orbit about Earth, with an orbital period of about ninety minutes.

Lithophage "Rock eater," from the Greek.

LMO Low Mars orbit. A near-circular orbit about Mars at an altitude of a few hundred kilometers.

Megaton The unit of explosive power, equivalent to the energy released by detonation of one million tons of TNT.

Mile An archaic measure of distance equivalent to 5,280 feet, 63,360 inches, 320 rods, 1,760 yards, 8 furlongs, 880 fathoms, 8,000 links, 7,040 spans, or 15,840 hands.

MSO Mars synchronous orbit. An orbit about Mars, between the orbits of Phobos and Deimos, with an orbital period of one Mars day.

Nanotechnology The near future technology of building, and building with, molecular-sized machines.

NASA National Aeronautics and Space Administration. An independent agency of the United States government charged with basic research on space, space technology, aviation, and the origin and distribution of life in space.

NEA Near-Earth asteroid. Any asteroid that crosses Earth's orbit or approaches within 1.3 AU of the Sun.

NEO Near-Earth object. Any NEA or Earth-crossing comet.

Ozone A molecule consisting of three oxygen atoms, O_3. An ozone layer in Earth's upper atmosphere protects the surface from dangerous ultraviolet radiation.

Regolith Shattered, crushed rock on a planetary surface; usually crushed by impacts.

S Asteroid Stony asteroid. Dominated by silicates of iron and magnesium, plagioclase feldspar, metallic iron-nickel alloy, and iron sulfide.

SDI(O) Strategic Defense Initiative (Organization). A military agency initiated by President Reagan to study technologies for defense against ballistic-missile attack. Developers of the Clementine spacecraft. SDIO was mocked as "Star Wars" by its Democratic opponents, who, upon winning the presidency, have continued to fund the program—under the alias of Ballistic Missile Defense Organization (BMDO).

Specific Impulse A measure of the performance of rocket engines; the number of seconds an engine will operate on one kilogram of fuel while producing a thrust of one kilogram. (Expressed in seconds.)

SPS Solar power satellite. A spacecraft that uses photovoltaic cells to convert sunlight into electricity, then beams that power down to Earth as microwave energy.

Washington A delightful city that combines the best of southern efficiency and northern hospitality.

INDEX

highly eccentric Earth orbit (HEEO), 125, 126, 132–135, 160, 242
highly eccentric Mars orbit (HEMO), 160
Hilda asteroid family, 188
Hiroshima, 83
Hirsch Prize, 22
Hitler, Adolph, 23
Hokusai, 256
Homer, 82, 189
Homo sapiens martiensis, 172
human-computer interfaces, 248
human inertia, 241
Hungaria asteroid family, 188
Huntsville, Alabama, 24
Huygens, Christiaan, 14
Huygens spacecraft, 52
hydrazine, 166–168
hydroelectric power, 217, 224
hydrogen in lunar regolith, 64
hydrogen-oxygen rocket engine, 21, 63, 103–107, 164, 165, 241, 242
hydrogen peroxide, 170
hydroponic agriculture, 231

ilmenite, 45, 47, 48, 64, 65, 69, 70, 103, 137–139
ilmenite reduction, 64, 69, 70, 72
impact erosion, 49
impact velocity, 91
importation to Earth, 111, 114
increasing-sum games, 235
India, 221, 222
induced radioactivity, 220
intelligence test, 233
Intercontinental ballistic missile (ICBM), 24, 25, 42
Intermediate-range ballistic missile (IRBM), 24
International Thermonuclear Experimental Reactor (ITER), 136, 246

interstellar probes, 242
interstellar travel, 26, 246, 255
ion engines, 245
Iran, 221, 222
Iraq, 221, 222
iron-nickel alloy, 48, 49, 89, 191
iron meteorites, 88, 91–93, 95, 112, 191
Israel, 222
Italy 222
Izhevskoye, 19

Japan, 42, 136, 222
Japanese unmanned lunar spacecraft, 42
Jet Propulsion Laboratory, 85, 161
Johnson Space Center, 121, 257
Journal of the British Interplanetary Society, 25, 179
Jovian outer satellites, 243
Julius Caesar, 15
Juno II booster, 36
Jupiter (planet) 10, 52, 97, 199, 205–208, 211
Jupiter perturbations of asteroid orbits, 96, 189, 190
Jupiter swingbys, 195
Jupiter IRBM, 36

Kaluga, Russia, 20
Kargel, Jeffrey, 114, 230
Kazakhstan, 40, 163
Kennedy, John F., 36, 37, 158
Kepler (lunar crater), 31
KGB (Soviet Union's Committee on State Security), 163
Khrushchev, Nikita S., 34–36
Kitt Peak National Observatory, 80
Komsomolskaya Pravda, 181
Korolev, Sergei P., 20, 34
KREEP basalt, 46
Kuiper Belt, 259
Kulcinski, Gerald, 135, 137

return delta V, 123
robots, 249
Rub al Khali, 226
Russia, 163, 218, 219
rutile, 72
Ryazan Province, Russia, 19

S-type asteroids, 94, 95, 111, 191
Sabatier reactor, 168, 169
Sahara Desert, 226
St. Petersburg (Russia) Society of
    Physics and Chemistry, 19
Salyut space stations, 118
Santorius, J. F., 211
satellite-based cellular phones, 250
Saturn (planet), 10, 97, 205–208, 211
Saturn swingbys, 195
Saturn (rocket boosters), 27, 39, 40,
    41, 55
self-replicating machines, 252
Shankara, 256
Shoemaker, Eugene, 79
short-period (periodic) comets, 97
Siberia, 218, 230
Singer, Robert, 45
Sinus Iridum, 32
skyhook, 180–182, 241, 242
Smith, E. E. "Doc," 186
Smithsonian Institution, 21
soft coal, 218
solar cells, 9, 67, 140, 225
solar collector, 66, 140
solar flares, 59, 102
solar power 11, 20, 26, 197, 198, 217,
    225–227, 256
solar power on the Moon, 66
solar power satellites (SPS), 131–134,
    141, 226, 227, 246
solar system, 5
solar thermal propulsion, 104–108,
    179, 242
solar wind gases, 47–49, 64, 70, 139,
    140, 204, 205

South Africa, 222, 230
Soviet ICBM program, 24, 25, 34, 35
Soviet manned lunar flyby program, 40,
    41, 58
Soviet Union, 163
Soyuz program, 39, 40, 118
space age, 5
space agriculture, 20, 61, 62
space habitats, 26, 194
space resources, 5, 25–27, 235
Space Shuttle, 6, 55, 56, 118, 122
space station, 110, 118–123
space station orbit: see low Earth orbit
space travel, 18, 20
Spacewatch, 7, 79, 80, 83
specific impulse, 104–107, 169
SPS (Solar Power Satellites), 8, 9
Sputnik 1, 25, 34–36
Sputnik 2, 35, 36
Sputnik 3, 36
Stalin, Joseph, 25
Stapledon, Olaf, 26, 255
steady state, 228
steam rocket: see solar thermal
    propulsion; nuclear thermal
    propulsion
stony-iron meteorites, 88
stony meteorites: see chondrites;
    achondrites
storable propellants, 109, 164, 167, 169
Strategic Defense Initiative (SDI), 42
strategic materials, 8
structural metals, 8
"summer exudation theory" of meteors,
    16
Surveyor program, 39
Sverdlovsk, 35
Sweden, 222

telepresence, 250, 251
telescope, 13, 93
terraforming, 172, 194, 195, 199
terrorists, 220

Made in the USA
San Bernardino, CA
02 February 2013